"*Mismatch* is a powerful read that not only has the potential to change the way we approach design but also serves as a strong check to our ingrained assumptions about how and why people move, act, speak, and interact (or don't)."

—*Gray Magazine*

"Designing for inclusion is not a feel-good sideline. Holmes shows how inclusion can be a source of innovation and growth, especially for digital technologies. It can be a catalyst for creativity and a boost for the bottom line as a customer base expands. And each time we remedy a mismatched interaction, we create an opportunity for more people to contribute to society in meaningful ways."

—*800-CEO-READ*

"One-of-a-kind ... Take[s] inclusive design out of an academic setting and into the working world. Nobody really wants to exclude people from their designs and this book shows you how you can avoid doing that."

—*Fast Company*

"Design shapes our human experience. As software changes our world, inclusive, thoughtful design will be even more important. In this absorbing and important book, Kat Holmes lays out clear steps and the role we can each play to overcome bias and create inclusive design."

—Victoria A. Espinel,
CEO, BSA | The Software Alliance

"Kat Holmes's approachable book calls to tech industry leaders and designers to create inclusion by making a world that invites all of us to participate and benefits everyone. Designing for the future, Holmes convinces, requires designing for human diversity. *Mismatch* is a manifesto, a primer, and a rousing invitation for everyone in the design and production process to become inclusion experts who will collectively make a better, more effective, and more just world for us to share."

—Rosemarie Garland-Thomson,
Professor of English and Bioethics, Emory University;
author of *Extraordinary Bodies* and *Staring: How We Look*

"Design at its best gladdens our lives and shapes the way we experience the world—but it can limit that experience, too. Counters may be too elevated for someone in a wheelchair, and playgrounds built solely for able-bodied children a trial for those who can neither run nor climb. Even a magical faucet, built to flow at the wave of a hand, is ill suited for the blind. In *Mismatch*, a crusading and important look at how design can frustrate and even alienate users, Kat Holmes compels us to be more inclusive. She shows us that design requires not only ingenuity but also humility. It must solve problems, yes, but it must also work with those excluded to reimagine and improve their experiences."

—Caroline Baumann,
Director, Cooper Hewitt, Smithsonian Design Museum

MISMATCH

How Inclusion Shapes Design

KAT HOLMES

The MIT Press
Cambridge, Massachusetts
London, England

First MIT Press paperback edition, 2020

This book was set in Scala and Scala Sans by Scribe Inc. Printed and bound in the United States of America.

Library of Congress Cataloging-in-Publication Data

Names: Holmes, Kat, author. | Maeda, John, writer of foreword.
Title: Mismatch : how inclusion shapes design / Kat Holmes ; foreword by John Maeda.
Description: Cambridge, MA : The MIT Press, 2018. | Series: Simplicity : design, technology, business, life | Includes bibliographical references and index.
Identifiers: LCCN 2018008099 | ISBN 9780262038881 (hardcover : alk. paper), 9780262539487 (paperback)
Subjects: LCSH: Design—Anthropological aspects. | Social integration. | Marginality, Social.
Classification: LCC NK1520 .H64 2018 | DDC 745.4—dc23 LC record available at https://lccn.loc.gov/2018008099

10 9 8

For Scarlet, Sophia, and Don

Together, we are home.

CONTENTS

FOREWORD

When I started to blog about the topic of design in the early 2000s, it was a time when there were still DVD players and the iPhone hadn't emerged yet. The cloud was just starting to cover the horizon of the sky-of-user-experience, and technology was just starting to make us happier and yet unhappier too. There needed to be a way to connect how technologists make products with the way pre-technologies were crafted by design—and thus the Laws of Simplicity (LoS) were born.

Luckily, I learned at an early enough age that I really don't have all the answers. So I have been actively looking these past few years for ways to think of design in the future by searching out the rising lights. A chance search hit on the Internet led me to the work of Kat Holmes on "inclusive design." This was back when she was at Microsoft—and frankly back then, Microsoft wasn't the first name you'd think of when you wanted to imagine the future of design. But, I thought, "This is it!"

Because, spending time in Silicon Valley and working across hundreds of technology startups as a partner at venture capital firm Kleiner Perkins, I intuitively felt that Kat had the solution to something truly important. Inclusive design was what was missing in the tech products being shipped to millions—ultimately leading to a fundamental mismatch between what people needed and what the techies in Silicon Valley were shipping out to them. I firmly believe that Kat holds the key to addressing the unfortunate set of common biases in how we make products in tech: for "ourselves" as a representative sample of the people in the world. So I reached out to her with a cold email and have been grateful ever since that she returned that first email message.

Kat Holmes brings us the right message about design, at just the right time. Her message isn't one of simplicity at all—she forces us to think about the complexity. According to how Kat sees the world, there can be no simplicity unless we understand the complexities of how and why products get built today. As you may recall, the 5th Law of Simplicity says:

Simplicity and complexity need each other.

Designing for simplicity tomorrow will be impossible unless we make the effort to understand the underlying complexities of how we design today. If we don't, we'll only create more mismatches. We'll create experiences that are simple for people like ourselves, only.

We need to ask the difficult question of who gets to make the products that we use today—because it ties directly into what gets made. This is the central question that Kat helps us wade into, with tact, theory, and concrete actionable advice for how to navigate this new way to design that is essential for any product maker out there.

As Kat says, "For better or worse, the people who design the touchpoints of society determine who can participate and who's left out. Often unwittingly." And, "If design is the source of mismatches and exclusion, can it also be the remedy? Yes. But it takes work."

Good luck in doing that work. I'm doing that work right now too.

John Maeda
Lexington, Massachusetts
March 2018

PREFACE TO THE PAPERBACK EDITION

We launched *Mismatch* in Detroit in September 2018 at a workshop hosted by the College for Creative Studies. More than three hundred students and professionals attended. I was most nervous about the four dozen high schoolers in the crowd. What would they think? What were the hardest questions they might ask, and how would they respond when I didn't have answers? Tiffany Brown, who co-led the event, was unfazed. She knew better. Those students *made* the workshop. They needed no invitation beyond this question: "What does exclusion mean to you?" It was up to the rest of us to listen and be attentive to their ideas.

When you deliver a book into the world, you send with it a generous amount of faith in the unknown. I'd imagine most authors, especially first-time authors, have moments of anxiety about how readers will react to their work. I certainly did. Inclusive design is so multifaceted and nuanced that there is always great potential for the discussion around *Mismatch* to fracture into factions.

I placed a high degree of faith in exclusion as a starting point for inclusion. I hoped it could be a unifying thread to mitigate the fractures between people and design. *Mismatch* is largely about learning to recognize exclusion in the objects we create and the environments we inhabit. What I didn't expect was the way shared experiences with exclusion could connect people to one another.

In the book's first year, I traveled to discuss inclusion with teams across sectors, including all kinds of digital and physical products, children's literacy, banking, housing development, recreational playscapes, and academia. From Seattle to Atlanta, Los Angeles to London, San Francisco to Singapore, people have shared how exclusion has shaped their own experiences and how *Mismatch* inspired a sense of belonging or a bold new action.

Such glimpses into the lives of so many people have been transformative for me. In the words of author, artist, and activist Cleo Wade, "No one's day is what you think it is. Be extra loving if you can."

Thank you to those who have shared how *Mismatch* resonates. Thank you for the countless pictures of *Mismatch* resting somewhere in your home, your favorite cafe, or the shelf at your local library. I'm particularly proud of a collection of pictures of cats sleeping on top of *Mismatch*. This little book has traveled much farther than I could ever have imagined.

My intention has always been to help start conversations that unlock inclusion. How an organization builds an inclusive design practice depends on too many factors to offer just one solution. So the answer is simply to start.

If *Mismatch* has in any way guided you to a new understanding or helped you broach the subject of inclusive design with your organization, thank you for taking these pages to heart and putting them into action. If you are new to *Mismatch*, welcome. Thank you for joining the conversation. I look forward to watching you build a more inclusive world.

Kat Holmes
October 2019

ACKNOWLEDGMENTS

Soon after I agreed to write *Mismatch*, my house flooded. Fifty thousand gallons of water and thirty-five thousand words later, it's all a bit of blur. These pages were written nomadically: in airplane seats, hotel lobbies, coffee shops, kitchen tables, library corners, and even on the front steps of several porches. I'm deeply grateful to all the people who shared their homes, meals, and friendship with my family during this time. In particular, the Woodman, Wither-Wollersheim, and Grimes families, my mother, Sharon Tangney, and her husband Steve. You created the safe spaces that made this book possible.

John Maeda, thank you for encouraging me to write this book and making it seem like the obvious thing to do. You've shown me what it means to open doors for others and I promise to pay it forward.

Bob Prior, your trust and creative partnership were a first-time author's dream. Thank you, and the MIT Press team, for the great care you brought to the nuanced topics in this book.

To the inclusive design leaders who contributed words to these pages: Tiffany Brown, John Porter, Victor Pineda, Sophia Holmes, Swetha Machanavajhala, Margaret Burnett, and Jutta Treviranus—thank you for sharing your expertise and stories.

The Airlift crew, thank you for creating a cover illustration that perfectly captures the essence of this book.

Karen Chappelle, your thoughtful and playful illustrations bring an important dimension to the ideas in each chapter.

Chuck Mosher, you coached me through the hurdles of writing. You endured my messiest thoughts to carve out my key points. Thanks for loving me that much.

Molly McCue, thanks for your eloquent feedback and being my writing soulmate.

Rosemarie Garland-Thomson, our long chats on misfits and mismatches helped me fit disparate pieces into one puzzle. And thank you for showing me how to write with Dragon when my words were in my mouth, not my hands.

Seema Sairam, Patrick Corrigan, Hsiao-Ching Chou, Sarah Morris, and Kris Woolery, thanks for your honest edits and remarkable friendship.

Irada Sadykhova, your compassion and clarity are woven throughout this book. Thank you for constantly challenging me, and a generation of inclusive leaders, to expect more of ourselves and each other.

To the extended community of inclusive design leaders and enthusiasts, especially at Microsoft, thank you for your partnership and support.

And finally, to my partner, Don, the rock that our family is built upon. When I decided to write a book, you didn't even flinch. While I wrote, you shifted entire planets around me so I could focus. Remember when I made you read scores of disparate pages, over and over again, and then accused you of not *really* having read the book? Thanks for thinking that was funny. Our life is made possible by all that you create.

1 WELCOME

Mismatches make us misfits.

Where did you love to play as a child? Maybe it was a tree near your home. A video game where you battled your way through new worlds. Or, like me, a fort you built out of boxes and blankets.

I've been talking about playgrounds a lot lately. I've been sliding on a lot of slides and trying out a lot of swings. I've listened to people talk about why they came to play and wondered who wasn't able to join in. All this may seem odd for an adult who designs technology for a living, but here's why it matters: Designing for inclusion starts with recognizing exclusion.

A playground is a perfect microcosm for learning how to start. Think back to the objects and people that occupied your play space. Try to remember what worked well for you. It's likely there were moments when you were happy to play alone and other moments when you played together with many children. Can you describe what it was that made those spaces inclusive?

It's also likely there were moments when you felt left out, either because there was an object that you couldn't use, or because you were ostracized by the people around you. What was it that made these spaces exclusionary?

The objects and people around us influence our ability to participate. Not just when playing on a playground, but in all aspects of society. Our cities, workplaces, technologies, even our interactions with each other are touchpoints for accessing the world around us.

When we meet those access points, sometimes we can interact with them easily and sometimes we can't. When we can't interact with ease, many of us will try to adapt ourselves to make the interaction work. There are also times when no degree of creativity will make it possible to use a solution that simply doesn't fit a person's body or mind.

Examples of this are all around us. It's the reason why a child climbs onto a counter to wash their hands at a sink. It's why people are left searching for instructions on how to navigate a software application when it's updated with new features. Anyone who's tried to order lunch off a menu that's written in a language they don't understand is in the middle of a mismatched interaction.

This is the power of mismatches. They make aspects of society accessible to some people, but not all people.

Mismatches are barriers to interacting with the world around us. They are a byproduct of how our world is designed.

Mismatches are the building blocks of exclusion. They can feel like little moments of exasperation when a technology product doesn't work the way we think it should. Or they can feel like running into a locked door marked with a big sign that says "keep out." Both hurt.

In this book we'll take a deep dive into how inclusion can be a source of innovation and growth, especially for digital technologies.

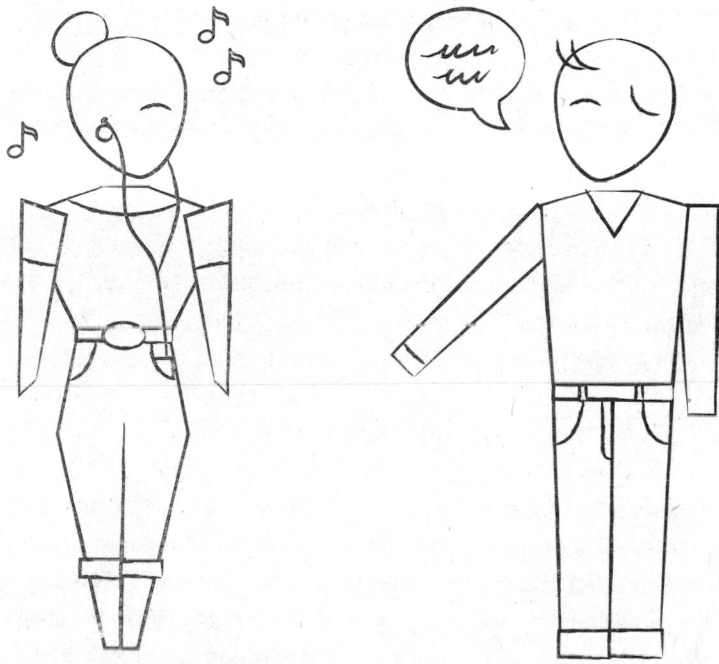

Figure 1.1
Human-to-human interactions are full of mismatches, but people are able to adapt themselves in an effort to connect with each other.

It can be a catalyst for creativity and an economic imperative. And we'll contend with a central challenge: is it even possible to design for all human diversity?

I've built a career promoting inclusion through design methods, also known as inclusive design. Like many people, I initially took inclusion at face value as a good thing. Yet I also found that people rarely made it a consistent priority. I wanted to understand why.

There are many challenges that stand in the way of inclusion, the sneakiest of which are sympathy and pity. Treating inclusion as a benevolent mission increases the separation between people. Believing that it should prevail simply because it's the right thing to do is the fastest way to undermine its progress. To its own detriment, inclusion is often categorized as a feel-good activity.

With this in mind, we will test our assumptions about inclusion and how it shows up in the world around us. We'll explore the reasons why our society perpetuates exclusion and new principles for shifting that cycle toward inclusive growth.

MISFITS BY DESIGN

What happens when a designed object rejects us? A door that won't open. A transit system that won't service our neighborhood. A computer mouse that doesn't work for people who are left-handed. A touchscreen payment system at a grocery store that only works for people who read English phrases, have 20/20 vision, and use a credit card.

When we're excluded by these designs, how does it shape our sense of belonging in the world? This question led me from playgrounds to computer systems, from Detroit public housing to virtual gaming worlds.

Ask a hundred people what inclusion means and you'll get a hundred different answers. Ask them what it means to be excluded and the answer will be uniformly clear: It's when you're left out.

Imagine children climbing on a playground. How are they climbing? Are they on a ladder, stairs, ramps, ropes, boulders, or maybe a tree? Now imagine who designs the features of that playground and the assumptions they make about the people who will play in it.

As you might expect, many playground designers are extraordinary advocates for inclusive spaces.

One cloudy San Francisco morning, while we were interviewing Susan Goltsman, she led our team through a park that she designed, pointing out the features that make it inclusive. A gently sloping ramp that reaches the highest lookout points. A harmoniously ringing gamelan, an Indonesian instrument on which "you can't play a bad tune."

A frenzy of children of various ages and abilities play together throughout the park. Their shouting and laughter are hard to compete with. With her feet in the sand, next to a giant sculpture of a sea turtle, Susan revealed the most important aspect of her design process:

> We interviewed kids with varying levels of disability, and the more severe the disability, the more vicarious the play. So the child who could not move very much was playing full-on in their brain, using other kids out on that play area to play through. So access means a lot of different things to a lot of different people.

Goltsman was a founding principal of the design firm Moore, Iacofano, Goltsman (MIG). Her influence extended far beyond childhood play spaces. One of the early pioneers in inclusive design, her contributions to policy and standards have influenced many major North American cities.[1]

An inclusive environment is far more than the shape of its doors, chairs, and rampways. It also considers the psychological and emotional impact on people. In partnering with Goltsman I learned that what's true for a playground is true for all human habitats, including the online world.

From a young age, we test the waters of acceptance by asking "can I play?" The response to this question can make our hearts soar or crush us. Over our lives we learn how to ask more subtly or we simply stop asking. Sometimes we push forward, regardless of rejection, to prove ourselves.

Core elements of our identities are formed by our encounters with inclusion and exclusion. We decide where we belong and where we're outsiders. It shapes our sense of value and what we believe we can contribute. Exclusion, and the social rejection that often accompanies it, are universal human experiences. We all know how it feels when we don't fit in.

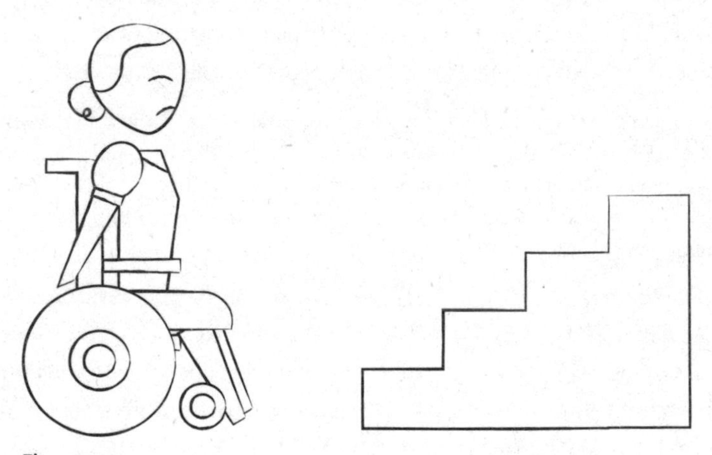

Figure 1.2
Mismatches between people and objects, physical or digital, happen when the object doesn't fit a person's needs. People often have to adapt themselves to make an object work.

For better or worse, the people who design the touchpoints of society determine who can participate and who's left out. Often unwittingly. A cycle of exclusion permeates our society. It hinders economic growth and undermines business success. It harms our collective and individual well-being. Design shapes our ability to access, participate in, and contribute to the world.

If design is the source of mismatches and exclusion, can it also be the remedy? Yes. But it takes work.

We must broaden our definition of design and designers. We must test our assumptions about human beings. We must wonder "who am I excluding?" and allow the answers to change our solutions.

Above all, we must be willing to acknowledge how much we don't know about inclusion. No one is an expert in inclusion in all areas of life. We are naturally better at exclusion, for reasons that we'll explore in the coming chapters. Knowing this, we can find better ways to forge ahead.

FEARS AND OPPORTUNITIES

There is a growing interest in making inclusion a positive goal for companies, teams, and products. The first actions for reaching that goal can be the most challenging. As with any expertise, inclusion is a skill that's developed with practice over time.

Where in life do we learn inclusive skills? In my education as an engineer, designer, and citizen I never formally learned about inclusion or exclusion. Accessibility, sociology, and civil rights weren't required curricula for learning how to build technology.

As I grew in my career as a technologist, I noticed a void of information on how to practice inclusive design for digital technologies. Most examples of universal design applied to sidewalk curb cuts and kitchen utensils. It was unclear how to achieve similar outcomes in the design of digital technology. In search of guidance, I realized that many people had the same question: where do I start?

Today, my answer to this question is always the same. There are many misconceptions about inclusion. It's important to know what you're getting into. These three fears of inclusion will likely strike you at some point. If so, you're not alone. But from each of them grows an insight into the nature of inclusion.

1. Inclusion isn't nice

One of the most common fears related to inclusion is a fear of using the wrong words. Many leaders would rather avoid the topic than look bad or offend someone. There isn't a robust lexicon for inclusion. There are many different interpretations of the word "inclusion," but very little guidance on what exactly this word means.

As a result, it can be easy to mistake nice words for good intentions. A person or company that offers the right language might not take meaningful action. It can also be easy to chastise someone who is committed to inclusion but who uses the wrong words. There's a strange irony when a group of people who are passionate about inclusion ostracize a person for saying the wrong thing. This is common across different languages and cultures.

That said, people sometimes say hateful things—and fully mean it. These words can reflect their true motivations. Often, they can be aimed at hurting groups of people who already experience the greatest amount of exclusion. This dangerous behavior is not addressed in this book. Instead, we will focus on the types of exclusion that emerge from being new to the topic and from unchecked exclusionary habits.

Inclusion isn't nice. It's challenging the status quo and fighting for hard-won victories. The opportunity is to be clear and rigorously improve our lexicon for inclusion. We can work on clarifying what we mean and why we care. We can create better resources through education and awareness. Asking questions, and then simply listening, is often the most courageous way to start.

Words hold a power to facilitate or freeze progress toward inclusion. Without a shared language, teams struggle to produce tangible results. The topic can often become emotionally charged if personal biases and pain dominate the conversation. Building a better vocabulary for inclusion starts with improving on the limited one that exists today. Sometimes we will use words that hurt people. What matters most is what we do next.

2. Inclusion is imperfect

The second fear is of getting it wrong. And you likely will, at first. You'll likely never achieve a perfectly universal solution that works for everyone in every situation. A common concern of designers is being forced to create a lowest-common-denominator design. Trying to please everyone is good for no one.

The underlying challenge is the vast complexity of human diversity. There are endless nuances and considerations when designing for people. There is no single answer that suits everyone. Accessible solutions are always, inevitably, accessible to some but not all people. A bathroom stall designed for a person who uses a wheelchair is often inaccessible to someone who stands three and a half feet tall.

Inclusion is imperfect and requires humility. It's an opportunity to be curious and approach challenges with a desire to learn. It teaches us new ways to adapt our solutions to what people need, which is sometimes different from how a designer thought their solution would work. In this book we'll look for unifying threads to guide us as we design for human diversity.

3. Inclusion is ongoing

The third fear is scarcity. There are rarely enough talented people, time, and money to make a sudden sea change in inclusivity. Urgency, especially in growth-led businesses, is a constant pressure. The tradeoffs are never easy.

As a result, the work of inclusion is never done. It's like caring for your teeth. There is no finish line. No matter how well you clean your teeth today, over time they require more care. With inclusion, each time we create a new solution it requires careful attention in its initial design and maintenance over time.

This underscores the beauty of constraints. We can learn how to choose great design constraints, ones that incorporate perspectives we haven't yet considered. It's a skill that we can sustain indefinitely if we build it into *how* we work, embedded into the entire process of creating solutions.

Inclusion is ongoing and in search of a better vocabulary. By association, so is this book. In writing it, I had to constantly remind myself that no one has all the answers. As a reader, I invite you to remember this as well.

In fact, I have a sneaking suspicion that it might not be possible to definitively design this elusive thing we call inclusion. Exclusion, conversely, is recognizable. It's measurable and tangible. When someone is excluded they know it unequivocally. The experience has an emotional and functional impact on them.

Perhaps, instead, all we can do is recognize and remedy the mismatched interactions in our world. The concrete nature of exclusion gives us something we can deconstruct. With our fears and imperfections in tow, exclusion is where we'll start.

WHY YOU, WHY NOW?

This book isn't an argument that everyone should be inclusive all the time. It's a case for why we should take responsibility for inclusion as a matter intentional choice, rather than risking an unintentional harm. Can we understand the exclusion created by our solutions *before* we release them into the world, and design something better?

Exclusion isn't inherently bad, nor inclusion inherently good. But in a society that sets goals such as a constitutional promise of equal rights and opportunity, barriers to that equality are problematic. For groups with capitalistic motivations, exclusion hinders business growth. A mismatch haunts any designer or technologist who aims to create great solutions but realizes just how much people struggle to successfully use their design.

These factors are amplified by the digital age. Technologies are permeating our public and private spaces. The modern marketer, engineer, or designer is expected to build solutions that reach millions if not billions of people. At that scale, one small exclusionary misstep can have an amplifying negative effect. Conversely, one small change toward inclusion can benefit many people in a positive way.

With respect to business justifications for inclusive design, there are four key categories that we'll explore in the stories ahead, with a review of each one in chapter 8:

- Customer engagement and contribution,
- Growing a larger customer base,

- Innovation and differentiation,
- Avoiding the high cost of retrofitting inclusion.

There are also concrete social benefits to inclusion. Each time we remedy a mismatched interaction, we open an opportunity for more people to contribute to society in meaningful ways. This, in turn, changes who can participate in building our world.

Designing for human diversity might be the key to our collective future. It's going to take a great diversity of talent, working together, to address the challenges we face in the 21st century: climate change, urbanization, mass migration, increased longevity and aging populations, early childhood development, social isolation, education, and caring for the most vulnerable among us in an ever-widening gap of economic disparity. You never know where, or who, a great solution will come from.

Already there are inclusive solutions quietly at work in our world. They are the early examples by which to measure inclusive outcomes. Their features, and the people who created them, share common threads which are captured here as three inclusive design principles. These will reappear in the coming chapters.

Recognize exclusion.

Learn from diversity.

Solve for one, extend to many.

Figure 1.3

- *Recognize exclusion.* Exclusion happens when we solve problems using our own biases.
- *Learn from human diversity.* Human beings are the real experts in adapting to diversity.
- *Solve for one, extend to many.* Focus on what's universally important to all humans.

These principles are derived from partnerships with inclusive design leaders, a heritage of successful innovations, and thousands of hours of applying them to product development. To include is a verb. In turn, these principles are also action-oriented.

In business and technology, we commonly look to leaders for guidance on how to be successful in new areas of expertise. I'm often asked for names of companies that are leading examples of inclusion. It's debatable whether any company is leading, yet. Many are talking about it, most are still early in their journey toward improvement. When it comes to inclusion, companies, and their leaders, have a lot to learn from a very specific kind of leader. It's not the prominent executive at a high-profile company. Not the faces that grace the covers of industry magazines. Not the people with the highest social media following.

We have the most to learn from leaders who've experienced great degrees of exclusion in their own lives.

Their expertise in exclusion means they can acutely recognize it in the world. It fuels their talent as problem solvers. This book follows leaders who are turning this expertise into action through design. They don't have all the answers, but they are finding better problems to solve. They are working through it, trying new angles, and navigating new paths that will benefit all of us.

With this in mind I've chosen specific stories of pioneers whose work on inclusion stems from their own exclusion. When they, and many more leaders with similar expertise, fill the ranks of the most visible positions in society, we'll know that all of us were good students of their work. Until then, I encourage you to seek out the excluded leaders in your community.

Often I meet new people, stories, and tools dedicated to advancing inclusion. I collect these at www.mismatch.design. These resources are a living companion to this book. I look forward to sharing them with you and evolving them with your input.

Whatever your reason for choosing this book, thank you for reading. Your contributions to building a more inclusive world will reach farther than you can imagine. You will find new ways to recognize and resolve mismatches in the world around you. In turn, you might be surprised by who shows up to play.

Welcome.

▪ ▪ ▪

TAKEAWAYS: MISMATCHES MAKE US MISFITS

- Inclusion is about challenging the status quo and fighting for hard-won victories.
- People's touchpoints with each other and with society are full of mismatched interactions. Design is a source of these mismatches, and can also be a remedy.
- Inclusion is ongoing, imperfect, and not nice.
- Exclusion isn't inherently negative, but it should at least be an intentional choice rather than an accidental harm.

2 SHUT IN, SHUT OUT

The games we play.

Imagine you work in an office with other people. One day, you arrive at work to discover a new rule is in effect. Maybe it's sent to you by the CEO of the company, or simply printed on a poster by the coffee machine:

> You can't say "you can't play." Effective immediately, if anyone wants to participate in your project you must agree to let them join. Yes, you will still be held accountable for the success of your work.

How would you react? Although the exact results might vary, there's a good chance that most adults might mirror the reactions you'd find with a group of young children: anger, defiance, and a few tears.

In her book *You Can't Say You Can't Play*, teacher Vivian Gussin Paley recounts what happened when she proposed this rule to her kindergarten class.

Before enacting the rule, she and her students speculated on what *might* happen. Their own reactions ranged from fear to excitement. Fear that their games would cease to be fun. Fear that an overwhelming number of people would want to join in, thus ruining the game. Or that unwanted people might intrude into the game. And the children who were most often excluded were excited about the protection the new rule would provide them.

Paley was inspired to introduce the rule after teaching countless students. Every year a few children in each class would be consistently excluded, sometimes to an extreme. As her former students became adults, they would recount stories of rejection as the most difficult times in their education. She created the rule as a way to study why this happened year over year, looking for ways to disrupt the pattern. Paley writes:

> Exclusion is written into the game of play. And play, as we know, will soon be the game of life. The children I teach are just emerging from life's deep wells of private perspective: babyhood and family.
>
> Then, along comes school. It is the first real exposure to the public arena. Children are required to share materials and teachers in a space that belongs to everyone.
>
> Equal participation is the cornerstone of most classrooms. This notion usually involves everything except free play, which is generally considered a private matter. Yet, in truth, free acceptance in play, partnerships, and teams is what matters most to any child.[1]

Paley's classroom of kindergarteners can help highlight why exclusion dominates many shared environments. Their unfiltered honesty gives us a clear view of why exclusion persists and how to evolve it. It starts with language.

THE CIRCLE

Many cultures have different approximations of the words "inclusion" and "exclusion," with unique origins and exact meanings. For the purposes of this book let's take a closer look specifically at the English words "include" and "exclude."

Both words are based on the Latin root *claudere,* which means "to close or shut." It represents a literal enclosure, but it also represents a mental model of separation. The most common image that comes to mind is a boundary created by a closed circle.

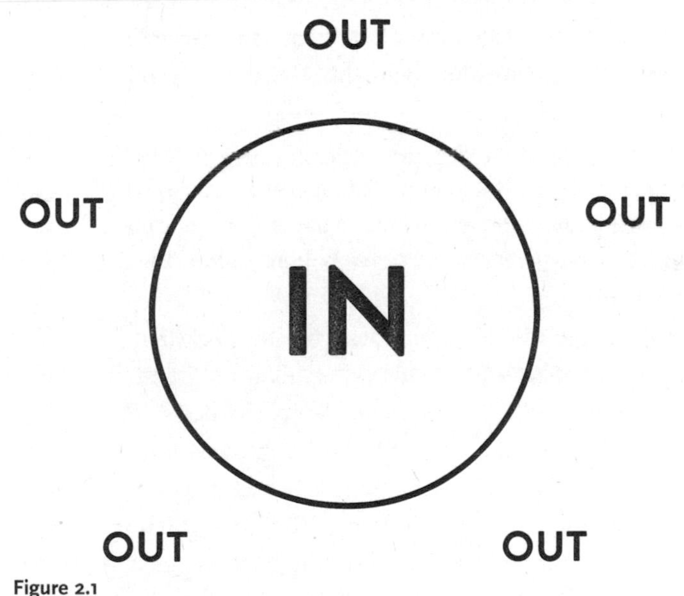

Figure 2.1
A shut-in-shut-out model of exclusion is centuries old and leads to a fixed way of thinking about inclusion.

Many modern-day societies use "inclusion" and "exclusion" to describe aspects of everyday life. Over time, such circular boundaries have been applied in new ways. They have been used to protect types of power and to separate people by gender identity, skin color, ability, language, religion, and other facets of human difference.

As a result, these societies also inherited the shut-in-shut-out mental model represented by these words.

When inclusion means so many different things to so many different people, understanding how to build it is far from self-evident. When companies talk about having an "inclusive culture," it's unlikely they want to shut people inside their walls. Inclusion is generally meant to express something more closely related to equity, empathy, access, or a sense of belonging. Somehow this little word has come to represent a vast world of good intentions, but if we can't describe exactly what it means, how can we start building it?

In the shut-in-shut-out model, what is the goal of inclusion? Is it for the people inside the circle to graciously allow people outside it to join them?

Are people who are shut out trying to break into the circle? Is the goal to obliterate the circle altogether and intermix freely in a utopian manner? The models that we use to describe the nature of exclusion inform how we think about solutions.

The circle encapsulates things that we want to protect, often power and possessions. The people we consider friends. The resources we believe are critical to our survival and success. As kids, we guard these things with phrases like:

"We already have enough people for this game."

"The game already started, we can't stop for you."

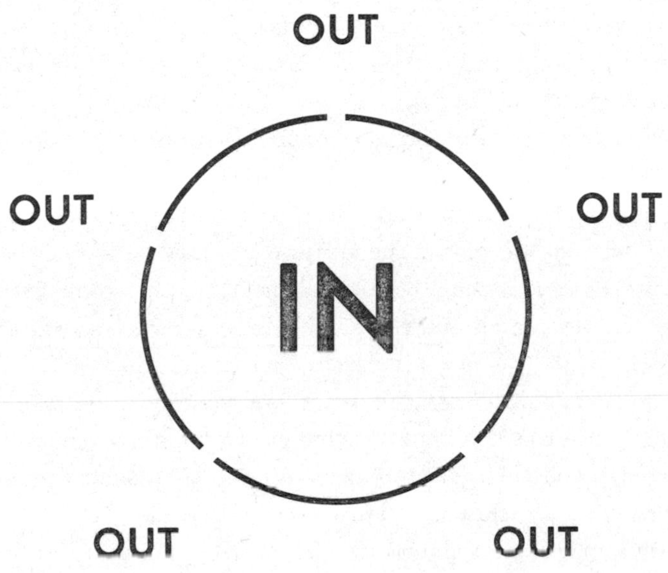

Figure 2.2
With a shut-in-shut-out model of exclusion, inclusion becomes a struggle between those in and those out.

"You don't have the right kind of toy to join us."

And yet, after the new rule was put into effect, Paley's students adapted their play with minimal conflict. The change was most difficult for a small number of children who consistently instigated games, set the rules, and enjoyed being the boss.

The children who had been most consistently excluded were no longer isolated. How they saw themselves and their contribution to the classroom shifted in positive ways. We'll explore this benefit in the next chapter along with the physiological effects of social rejection, such as physical pain and depression.

However, the subtlest and broadest benefit was that every child in the classroom gained new friendships, and the games they played got a lot more interesting.

When they could no longer exclude each other, they learned how to adapt their games. They also adapted the roles they were willing to play within a game. They tried on different identities. The kid who was always the villain could now be the newborn baby. The heroes could be the villains. The father could be the mother. Despite all initial concerns, their games were still fun to play.

These childhood fears ring true in our adult lives. We meet the same concerns when improving the inclusivity of our workplaces, products, and public environments. Paley's classroom experiment illustrates that exclusion isn't based on a fixed circle.

It's a cycle of our own making.

A FRAMEWORK FOR CHANGE

Unlike in Paley's classroom, we don't always have, or want, a higher authority posting rules on our walls to mandate our behavior. The good news is we each hold the power to make or break inclusion. The bad news is we don't always know that we have this power, or what to do with it.

We each know what it feels like to be excluded. Sometimes we can recognize when someone else is left out. Yet we often have a very difficult time anticipating how people might be excluded in the future.

This is what makes inclusion so relevant to design.

A design has an intent. It's meant for a purpose. The act of designing inherently requires thinking about the future ways that someone might use a solution. And it's successful only when the recipient of a design confirms it has achieved its purpose.

Inclusion complements design as a way to align what a solution *can be* with what a person *needs* it to be. This dynamic is best described by Dr. Victor Pineda, a leader in accessible urban design and co-founder of the Smart Cities Initiative:

> Inclusive design is about engaging with people that can be completely different than you, it stretches your imagination of what's possible. It has a trickle effect, it has a multiplier effect in that it changes those people and, in a sense, changes society.
>
> Because the fact that the designer thought about a wider group of people opens up for society to see these people that were once invisible.[2]

Dr. Pineda speaks from experience. He's one of the most well-traveled people you'll ever meet, having visited over seventy countries. As a person who uses a wheelchair he has expertise in navigating obstacles in public spaces. He has a combination of strategies, from personal assistants to assistive technologies, that are critical to each of his journeys.

He has an intimate understanding of how interrelated elements, from designed objects to city policies, can create and remove barriers to access. And what happens when they do:

> Designers, whether they're designing a school or software, hold the key to unlocking human potential. Because I want to give all I have to society. You can do that. You can change the rules of the game so the game includes me and includes my talents.[3]

What makes exclusion a cycle? It's the enduring notion that "we shape our tools and thereafter our tools shape us."[4] What we produce has an effect on society, which in turn shapes the next set of problems we aim to solve. A solution becomes a barrier when it's designed only for people with certain abilities. The brainpower and ingenuity of anyone who doesn't match that design are simply untapped. When we create new ways for people to contribute their talents, their contributions influence everyone.

Exclusion is also cyclical because it's constantly renewed by our choices. For example, changing the colors of a website could seem like a small adjustment. Or it could suddenly make that website inaccessible to nearly 8% of men and 0.4% of women who have color blindness.[5] The cycle is always in motion and shifts with each design decision.

We will use the framework of a cycle to examine how we perpetuate mismatched designs and how to shift toward inclusion. There are five elements to this cycle, and each of them is interrelated with the others.

- *Why we make.* We'll focus on the motivations that are innate to the problem solver.
- *Who makes the solution.* This is the problem solver, the person who is accountable for the success of a solution.
- *How we make.* These are the methods and resources employed by the problem solver.
- *Who uses it.* These are the assumptions that the problem solver makes about the people who will interact with, receive, or benefit from the solution.
- *What we make.* This is what the problem solver creates.

You might notice that the cycle doesn't include *when we make.* This is because exclusion can show itself at any moment in the development process. Conversely, the best time to start remedying exclusion is right now. It is most effective, and generally less expensive, to prioritize inclusion as early as possible and build inclusive solutions from the ground up.

Figure 2.3
The five elements that contribute to a cycle of exclusion.

But design rarely starts with a blank canvas. Nor does it follow a neat and sequential process. When inclusion is characterized as a separate category of activities reserved solely for the early stages of a design process, it can easily be deferred indefinitely. In short, right now is better than never.

That said, in chapter 8 we'll take a closer look at examples of inclusive design. Some of these are modifications of existing solutions. Others were created by applying inclusive design in the early stages of a new business or product. We don't need to tear down existing solutions to make inclusive ones. The cycle of exclusion is pervasive and ongoing. Similarly, a shift to inclusive design means we are constantly looking for and resolving mismatches through all stages of a development process.

As an individual or a member of a small team, you might have a high degree of control over all of the elements in this cycle. With practice, you can learn to recognize and remedy exclusion at your own pace.

Larger organizations often have a harder time coordinating the elements of this cycle. Sometimes the elements are isolated silos that are disconnected from one other. This issue is aggravated when leaders try to build inclusion through only one element of the cycle.

Leaders often focus on the single element that relates to their professional expertise. An engineer might focus on accessibility issues with the product they develop. A leader in human resources might focus on hiring practices. An information technology professional might focus on how to make communication tools work better between global teams that have different native languages.

The success of inclusion for each element is gated by the success of the other elements.

Hiring a new engineer whose first language is Mandarin and requiring them to use English-only tools could impede their ability to perform to their full potential. A team that consists of people with perfect eyesight might build a touchscreen interface for a camera without ever taking into consideration how it would work for someone with vision loss.

Whether by lack of awareness, siloed decisions, or simple neglect, it can be difficult for organizations to drive toward inclusion if they don't have a full picture of how their *existing* culture perpetuates exclusion. As a result, the default state for most organizations is a cycle of exclusion.

The power to change that cycle doesn't just belong to the person who starts the game, but to all who participate in it.

▪ ▪ ▪

TAKEAWAYS: THE GAMES WE PLAY

- The shut-in-shut-out model of inclusion can make it seem like exclusion is a fixed state, rather than an active choice.
- Reframing exclusion as a cycle helps us recognize the ways that our design choices can lead to mismatches.
- The five elements of the cycle of exclusion are interrelated. Each can be a useful place to start shifting toward inclusion.

3 THE CYCLE OF EXCLUSION

Why it's time to kick the habit.

What do a public bathroom and a smartphone have in common? From the standpoint of inclusion, they're similar. The cycle helps illustrate just how much.

Public bathrooms are vivid examples of exclusion. An "accessible" bathroom is most commonly one that meets basic building standards for wheelchair accessibility. Architectural features that create access for someone who uses a wheelchair are critical, and mandated by law in some countries. And yet, the next time you visit a public restroom, take a moment to consider who, exactly, it is accessible to.

Here's a cringeworthy example. There's an increase in toilets that can only be flushed by waving your hand over a sensor on the back of the toilet. It's a futuristic invention, complete with a friendly icon of a waving hand to demonstrate how to activate the toilet.

Unless, of course, you're unable to see the icon and are expecting some form of button or lever to flush. One of the last things anyone wants to do is reach around a public toilet in search of how to flush it.

Here are a few more perspectives to consider. Are the door locks and toilet seats reachable for someone who's under four feet tall? Or over seven feet tall? What physical features are required to access the sinks and faucets? Do the automated soap dispensers respond to people with a wide range of skin tones? Is it a safe space for people across a range of genders? How well does the space work for children and their parents? How well does it work for people with luggage? For a person with a broken ankle?

For a team that designs bathrooms these are important questions to ask. If someone on the team has experienced these kinds of exclusion, they likely have expertise that's helpful to creating a better solution. If a team doesn't have the experience, and isn't held accountable for finding a solution, they will likely build a solution that excludes.

For bathroom visitors, often the person who needs to use the solution can't simply choose not to use it. An exclusive design places a burden on people to find their own workarounds. A hands-free, sensor-enabled toilet leaves people searching the surface of a toilet for a handle to flush, or asking another person to help them. These can be dehumanizing experiences.

Public spaces, like bathrooms, serve universal human needs. They are access points for health, safety, education, economic opportunity, and Internet connectivity. It makes sense to have rigorous criteria that minimize exclusion in these spaces. But do designers need to be as mindful when designing technology? Absolutely.

Disability is critical to any conversation about exclusion. It touches everyone's life, eventually. Yet disability is commonly misunderstood as applying to only a marginal percentage of the human population. This is simply untrue.

According to the World Bank, one billion people, or 15% of the world's population, experience disability.[1] To put it another way, there are over 6.4 billion people who are temporarily able-bodied.

As we age, we all gain and lose abilities. Our abilities change through illness and injury. Eventually, we all are excluded by designs that don't fit our ever-changing bodies. We'll take a much closer look at this dynamic in chapter 7 as we challenge some outdated notions about what it means to be a normal human being.

Figure 3.1
Everyone gains and loses abilities over the course of their lifetime.

Consider how much of our world is visual. Large amounts of information are conveyed through computer screens, street signs, and various uses of light. An icon isn't just an icon. It's a way to communicate information. All people, regardless of their degree of eyesight, want to be able to access these kinds of information, whether in reading a lunch menu or sharing a quirky selfie with your best friend.

Touchscreens are making their way into many environments, often replacing human beings with an automated system. They are often the key to buying groceries, navigating transit stations, pumping gasoline into a car, and even completing classroom assignments. But touchscreens can make these spaces inaccessible to people who are unable to see or touch the screen.

These kinds of exclusion impact many people with varying degrees of sightedness. There are people who are born blind. People with one of many kinds of color blindness, or partial vision. There are people with light sensitivity, farsightedness, nearsightedness. Given a long enough life span, everyone loses some degree of their eyesight.

As our bodies change, we encounter more mismatched designs, simply because the designs that once worked well for us don't change with us. There are two key side effects that stem from being subjected to high degrees of exclusion: societal invisibility and the pain of rejection. Let's take a closer look at each one.

INVISIBLE BY OMISSION

Exclusion isn't evenly spread across social groups. Although we all experience it in our lives, some communities of people are ostracized in more ways and over longer periods of time.

A designer recently confessed to me that they would like to learn more about disability, but that the diversity topic that they cared most about was gender. Why would they feel the need to choose one and omit the other?

This is another kind of mismatch. How we categorize people often doesn't reflect how they really are: multifaceted. When we design, which facets of human beings are most relevant to our solutions? We'll explore this question in chapter 4.

Grouping people based on oversimplified categories like "female" or "disabled" or "elderly" can seem like a helpful shorthand when making business or design decisions. But there's one big problem:

Some categories of people are always last on the list of priorities, or wholly forgotten.

As Paley noticed with her students, the same few children were rejected year after year. They were "made to feel like strangers."[2] This phenomenon can happen to entire categories of people.

Disability is one such category. Exclusion has a profound impact on disability communities, and the consequences are rarely given the attention they deserve. The societal omission of disability runs deep. So deep that entire populations of people are virtually invisible in society.

For example, disability rights and history are rarely incorporated in school curricula. Accessibility is scarcely a required course for engineers and designers. Formative disability rights leaders and the movements they champion are seldom mentioned alongside other civil rights leaders. These omissions can further reinforce stereotypes and isolate people with disabilities from society.

Yet the influence of these leaders permeates the design of our world. Curb cuts are the quintessential example. A curb cut is the sloped transition from a sidewalk to a street that makes the crossing accessible to people who use wheelchairs.

Between 1970 and 1974 Telegraph Avenue in Berkeley became an extensively accessible route for wheelchairs, making it one of the first streets in the United States to do so. It was one culmination of changes sparked by Ed Roberts and leaders of the Independent Living Movement, in their fight for disability rights on the University of California campus. The design was evolved and refined over time to include adjustments for people who are blind, who had relied on the shape of the curb as a way to avoid accidentally walking into street traffic. Different textures were used to indicate the transition into a street, as many corners continue to do today. Curb cuts are also widely used by anyone who's pushing a baby stroller, towing a suitcase, or riding a bike.

Another example of societal invisibility is the low labor participation rates for people with disabilities. Just 35.2% of people with disabilities in the United States were employed in 2015.[3] This is compared to a 62% labor participation rate for people without disabilities in the same year.[4] Multiple sources have different statistics, but they all point to the same general truth.

There are many other significant gaps in equality for people with disabilities. Design, alone, isn't going to close all of these gaps. Yet designers, engineers, and business leaders can make progress toward equality with every design decision they make. And it starts with building a baseline understanding of the accessibility policies that impact their work.[5]

These examples reflect how exclusion impacts people with disabilities. Similar examples can be found for communities that are excluded based on ethnicity, gender, economic status, and more.

For designers, one important way to change invisibility is to seek out the perspectives of people who are, or risk being, the most excluded by a solution. Often, the people who carry the greatest burden of exclusion also have the greatest insight into how to shift design toward inclusion.

HEARTBREAKING DESIGN

If we could prove that exclusion causes physical pain, what would we change about the design of our classrooms? Our workplaces? Our technologies?

We have rules in schools and society against physically harming each other, but for some people being left out is treated as a fact of life. It's just part of the way the game was designed. It's fairly common for people to treat social rejection as a necessary part of learning to survive in the world of humans.

Yet multiple studies show that social rejection might manifest in our bodies in ways that approximate physical pain.

This is exactly what Ruth Thomas-Suh reveals in her film *Reject*. She features a community of researchers and academics working to understand the links between exclusion and pain.[6] They all point to a similar insight: rejection hurts.

More specifically, they found that the regions of the brain that "regulate the distress of social exclusion" were similar to the regions that regulate physical pain.[7] In other words, being socially rejected might directly affect our physical well-being.

This rejection has many consequences: anxiety, insecurity, anger, hostility, feelings of inadequacy, a sense of being out of control. Even the words that people use to describe social exclusion approximate physical pain. Hurt feelings. A broken heart.

What if an object rejects us? Is it as painful as being rejected by a person?

Dr. Kipling Williams was relaxing at a park one afternoon when a wayward frisbee landed in his lap. He stood up and tossed it back to two people who were playing frisbee. They started throwing the frisbee to Williams, including him in their game.

Suddenly, without a word, they stopped. They continued to play with each other, but Williams was no longer included. The experience stuck with him, with its feeling of dejection.

He quickly realized that he could recreate this experience in his laboratory at Purdue University to study the effects of exclusion. He made a simple computer simulation, called Cyberball, in which two characters on a screen tossed a virtual ball to the research participant, who was led to believe that the two other players were real people in another room. Then, suddenly, the two characters stopped tossing the ball to the research participant, effectively excluding them for the rest of the game.

People who participated in Williams's Cyberball studies typically reported a degree of sadness and anger that often accompanies rejection. This was expected.

What surprised Williams was that their anger increased once they learned that the two other players were controlled by a computer, not by real people. One person aptly described their reaction as, "You know that people will let you down, but computers aren't supposed to."

We expect people to be unfair. People are fallible and prone to faults. Yet we expect technology to be impartial. Maybe it doesn't always work the way it's supposed to, but we tend to believe that inanimate objects are largely unbiased.

That is, until we're locked out of accessing a building, product, or event that most people can access without difficulty. When the cycle of exclusion is in effect, the resulting designs are far from fair.

Not all exclusive designs are negative. They simply reflect a series of choices made by people who have the power to set the rules of a given design. Clothing made to fit certain body shapes. Special-edition products created to generate interest in a new business. An invitation-only birthday for an inner circle of friends. Sometimes creating limited access can have positive benefits.

The problem comes when there's a mismatch between the stated purpose of a design and the reality of who can use it.

When a solution is meant to serve any member of society and then doesn't, the effects of exclusion can be negative. The experience can feel like rejection from society itself. Especially in the shared physical and digital spaces where we learn, work, share, heal, advocate, create, and communicate.

There is a risk that exclusion will become more prevalent as technology moves into every area of our lives, because interactions that were once human-to-human are now facilitated by machines. Every human interaction that includes technology gains a wild card: who will it reject and who will it accept?

EXCLUSION HABITS

What leads to cycles of exclusion and why are they more prevalent than inclusion?

Every day, design teams make incorrect assumptions about people. They might presume that someone who's blind doesn't use a camera. Or that a person who's deaf doesn't use a music streaming service. Is it malicious intent? Not likely.

Let's consider the first element of the cycle: Why we make.

As we set out to solve a problem, let's imagine that most people have best intentions to create a beneficial solution. Their goal might be to solve a known issue for their existing customers. This is often in competition with other goals, like meeting a due date, looking good in front of someone in an authority position, or earning a reward for driving new business growth. In the crucible of competing goals, even the best intentions can get lost in the shuffle.

There can be little time to think. We might want to pursue our best intentions for inclusion, but there's a constant pressure to keep growing and moving quickly. Like a carousel on fast forward. In these moments, assumptions become a necessary shorthand.

And then something much more powerful takes the lead: habit.

Paley refers to this as the *habit* of exclusion. Without an explicit understanding of how exclusion works, our default habits—habits we formed in our early childhood—can dominate, creating a cycle of exclusion.

An exclusion habit is the belief that whoever starts the game also sets the rules of the game. We think we don't have power to change a game, so we abdicate our accountability. We keep repeating the same behaviors, over and over.

In short and simple games, it might be easy to call on the one who's responsible for changing the rules to make it more inclusive. Over time, games get more complex, leaders change, and we can forget who authored the original rules. In some organizations, the cultural behaviors were set a long time ago and the founders of that culture are long gone. Or we believe it's someone else's job to rewrite the rules, maybe the leaders in our business or community.

We also forget that those rules were initially written by human beings and can be rewritten. Those of us who are now playing the game have a responsibility to adapt it as needed. If we don't, we are accountable when someone's left out—not some leader from

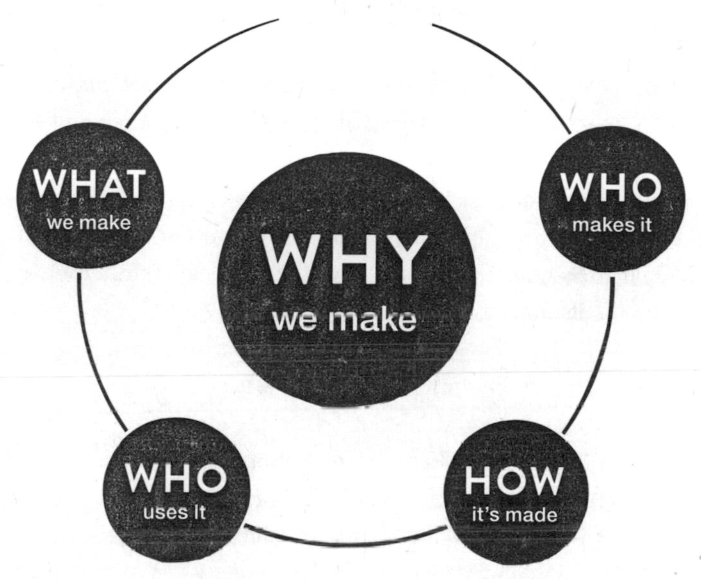

Figure 3.2
Exclusion habits are the reason why we make mismatched designs. They stem from deep-seated assumptions about the people who receive our designs.

the distant past. We can respect the intent of the game, but also adapt the rules to make it more inclusive.

An architect might create a building with a grand staircase leading to its front entrance on the basis of tradition or aesthetics. Meanwhile, people who use wheelchairs might be searching for back-alley entrances and convoluted hallways to access the building.

When making choices in the design of a solution, we might think "those are just the rules, I didn't make them up." It's easy to defer responsibility by claiming this is just how the world worked when we arrived.

This is why it's important to distinguish *why we make* solutions, especially when we work on behalf of companies and organizations. While anyone can start to shift their personal reasons for creating solutions, inclusion often needs to be made part of an organization's culture by people in the most senior leadership roles. If inclusion isn't explicitly part of that leadership, exclusion will be the default.

BUILDING NEW HABITS

Exclusion habits can be hard to break. But, like any habit, they can be changed over time with new practices to challenge our mindsets and behaviors. When we say that we're committed to inclusion, it's like declaring that we'll learn a new language. We start the next day full of enthusiasm and optimism, but become quickly aware of our huge gap in expertise.

Learning a new language can take planning, training, and determination. But above all, it means engaging with people who are native in the new language you want to learn.

To gain fluency, you will need to change aspects of your routine and adjust some the elements of your life to support your new goal. You might even relocate to a community where your new language is spoken every day. The same is true for building skills for inclusion.

These skills can be learned from people who interact with unwelcoming designs every day of their lives. They often have an intimate understanding of all the angles to consider. These are the designers, engineers, and leaders who have the greatest power to disrupt the cycle of exclusion.

Learning from these experts, we will identify the top exclusion habits for each element of the cycle. And we will highlight ways to shift the cycle toward inclusive design.

■ ■ ■

TAKEAWAYS: WHY IT'S TIME TO KICK THE HABIT

- Mismatched designs contribute to the societal invisibility of certain groups, like people with disabilities.
- When a designed object rejects a person, it can feel like social rejection and approximate physical pain.
- Exclusion habits stem from a belief that we can't change aspects of society that were originally set into motion by someone other than ourselves.

4 INCLUSIVE DESIGNERS

Building the skills to recognize and resolve mismatches.

John R. Porter's bedroom would make any die-hard gamer feel right at home. Among the personal computers, beyond the *Run Lola Run* movie poster, across from a MakerBot busily extruding plastic threads into an unidentifiable widget, is a large black pegboard. The kind of board my grandfather used for hanging hammers and wrenches in his garage.

Mounted on the pegboard, like hunting trophies, are dozens of video game controllers dating back decades. Among them is the 1977 nostalgia-inducing Atari Video Computer System controller, with its joystick and single red button. Hundreds of millions of people used this kind of controller to play the legendary game *Pong*. Next in line is the 1985 iconic block-shaped Nintendo Entertainment System controller with two buttons and a T-shaped directional pad that players used to maneuver through *Super Mario Bros*, one of the most popular video games of all time.

Symbiotically, video games and their controllers grew more complex in the 1990s as home video games grew in popularity. Porter's pegboard features the contoured dual grips of the 1994 Sony PlayStation gamepad and a few hefty Microsoft Xbox controllers released since 2001, adorned with buttons, sticks, triggers, and a directional pad.

Figure 4.1
Early designs for Atari, Nintendo, Sony's PlayStation, and Microsoft's Xbox game controllers.

Moving along what Porter calls his Wall of Exclusion, it's easy to notice how the controllers grew larger, heavier, and more complex over time. But one thing unites them all. They all require two hands to play.

These controllers are gateways to vast virtual worlds. In these worlds players achieve new skills, explore complex landscapes, and connect with each other. For many years, Porter admired these games from a distance, excluded from playing them by the shape of the controllers.

We first met when Porter was a design intern at Microsoft. He was working on his PhD while teaching at the University of Washington's Human Centered Design and Engineering program.

Better than most people, Porter can tell you exactly how to create an inclusive solution. Technology is very tightly integrated with everything he does. He uses a wheelchair to get around and assistive technologies to extend his own abilities. These technologies help Porter bridge the mismatched designs that he encounters when using everyday objects and spaces.

In a world where most technology is designed to be used with a keyboard, mouse, or touchscreen, Porter depends primarily on speech recognition. He uses his voice to interact with computers through a software program called Dragon, created by Nuance Communications, Inc. Dragon listens and follows his commands. By talking naturally to his computers Porter can create lesson plans, compose written communications, and participate in online gaming.

He's active in a community of gamers with disabilities that share elaborate techniques for how to hack together alternate ways to play their favorite games. Sometimes this is a hardware hack, like a switch controller that can be modified to work through head movement. And sometimes it's a software hack, like programming a sequence of game actions into one simple command. For example, just by saying "prepare for attack," a program can coordinate multiple characters to aim in the same direction at once, rather than having to align each one manually.

Porter ties his experience with gaming to everyday life, making insightful connections between play and inclusion. Here are some excerpts from a conversation with Porter to guide our exploration of the second element in the cycle of exclusion: *who makes it*.

Why did you build a Wall of Exclusion?

I keep these here to remind me of all the assumptions that we, as designers, make about people. The design of these products clearly signals that gaming is for some people, and not for others. A game controller says "This is for you" or "This is not for you." This is true for everything we design.

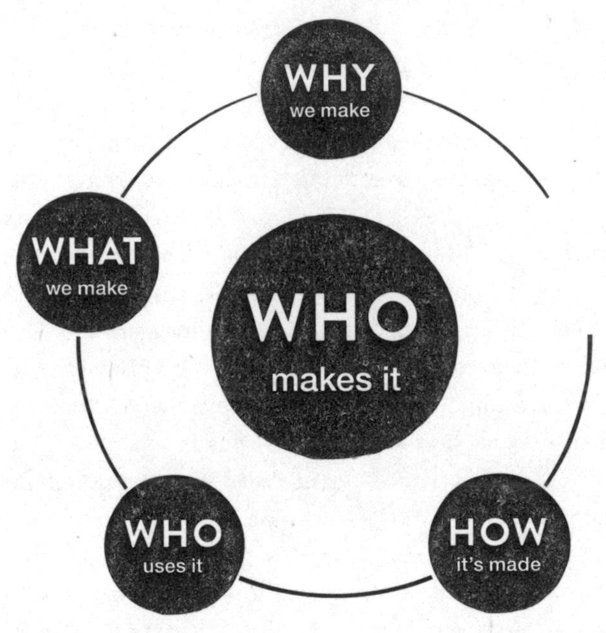

Figure 4.2
The people who make solutions hold a power to determine who is and who isn't able to participate.

What happens when designers make assumptions about people?

All of these games are based on a broad assumption that you'll be using your fingers and your hands to interact. And for me that is almost entirely moot. I have some physical mobility that I can use to move my wheelchair, but everything I do in the digital realm is mediated through other avenues of interaction, primarily speech.

Whenever I'm using technology I use speech to control it even though it was never designed to work for speech. It's not just that a game isn't optimized for my abilities. It was made without ever even considering the possibility that someone would need to interact with it in the ways that I do.

That places the onus on me to figure out all these workarounds. For gamers with disabilities, we have to spend as much time figuring out how to play a game as we do actually playing.

How does a designer make something inclusive?

Games that only allow a user to play in one way, that have a very prescriptive notion of who a player is, those tend to be the ones that are the least accessible. But games that allow more freedom and flexibility tend to be a lot more inclusive.

I often like to point out *World of Warcraft* as one of the really great inclusive games. There are players who don't have the motor ability to interact fast enough to engage in combat. But I know people with disabilities who have played this game for years. They have all of their crafting skills maxed out because for them, the game isn't about doing quests.

The game is economics.

The game is building a leather-working empire and making gear that people can buy from them. And I don't know if anyone at Blizzard Entertainment would say that they envisioned that as a viable play style. But nevertheless, they built an inclusive system.

What's the role of gaming in your life?

I still remember the day that my uncle Mike came over to give me *Final Fantasy VII* as a 12th birthday present. It was my favorite game, despite the fact that I'd never played it. I'd watched him play it, but was never able to join in. I was able to play games until I lost the ability to use game controllers around the age of 10. For me, gaming was becoming a spectator sport.

My uncle once spent an entire afternoon trying to tape little pieces of wood onto his controller to make it work for me. When the efforts were unsuccessful, I told him not to worry about it. *Final Fantasy* just wasn't for me. "Sure it is," he reassured me as he picked up the modified controller, "we just need to stop this thing from getting in the way."

I don't just play. I work to figure out how to play. It's figuring out how to participate in societal moments. Which is a burden sometimes. I think of it as a metagame that I play in all areas of my life. It's a puzzle to be solved and shared.

How is gaming shifting from exclusion to inclusion?

There is a natural inclination to use yourself as a shortcut to make assumptions about the people that you're designing for. And I think that's especially rampant in the world of game design.

Historically, the industry has been incredibly homogenous. Games have been created by people who feel like they are never going to be different than they are in that moment. Many of them feel like they are the only users who really need to be considered. And that's changing, which is good.

I don't think it's a coincidence that the increasing diversity amongst game designers is happening concurrently with the release of more inclusive games. I think those two things go together.

Porter also points out that for decades, games and consoles were made exclusively by large companies with massive technology requirements that took years to release. Only an elite group of designers worked for these companies.

Today, a growing community of gamers with disabilities, and organizations like AbleGamers, are pushing the creative boundaries of how people design and access games. Years of patience and sheer resourcefulness are opening up the benefits of gaming to a much wider audience.

Also, with multiple open ways to make and publish games, a more diverse group of people are creating more diverse content for people to play. A new generation of inclusive-minded designers and enthusiasts is gradually transforming the gaming industry.

Inclusive designers can emerge from traditional design disciplines. They also come from unexpected backgrounds. It's important to define the specific design skills that contribute to inclusion, so that more people can become practitioners. This is also why we must reimagine what it means to be a designer.

DECIDING WHO DESIGNS

The power of shifting *who makes* has had a similar effect in industries beyond gaming. More designers are focused on how to adapt objects to make them work for a diversity of people. Open-source tools enable more people to contribute to the design of everything from education to artificial intelligence. The cycle of exclusion shifts toward inclusion when more people can openly participate as designers.

Anyone who has ever solved a problem is, in a certain sense, a designer. The only real difference comes in how much ownership you take over the identity of yourself as a designer. You might be a designer if you say that it's not enough to design for yourself and you want to design experiences for other people too.

Consider the rigid ways that companies hire new employees. Many companies require candidates to complete an online application, an often-tedious process that requires specific language competencies, access to the Internet, and an ability to focus on detailed information for long periods of time.

In the tech industry, a common approach to interviews is for a candidate to meet sequentially with multiple people in a face-to-face verbal interview. The questions are designed to assess how likely a person would be to be successful in the role they're applying for. As the day wears on, a candidate needs to have a high degree of physical and mental stamina to endure the process.

In the end, the hiring process itself can send a clear signal that the job is "for you" or "not for you." Yet how many aspects of this process have a direct correlation to how well a person will really perform in a role over a multiyear career?

With this in mind, let's take a closer look at three skills of successful inclusive designers:

1. Identify ability biases and mismatched interactions between people and world.
2. Create a diversity of ways to participate in an experience.
3. Design for interdependence and bring complementary skills together.

ABILITY BIASES AND MISMATCHED INTERACTIONS

Many organizations create solutions for thousands if not millions of people. Creating solutions at such a large scale means hundreds, maybe thousands, of designers and engineers are working together on various aspects of a solution. When all those team members bring their own biases to the process, it can be challenging to make a solution that works well for all the people it is intended for. In fact, it can seem absolutely impossible.

The underlying challenge is human diversity.

A common starting point for teams is to focus on increasing the demographic diversity of their team members. Representation of diversity is important. However, changing representation doesn't necessarily change culture. Culture change can be hard and takes time. If we increase diversity of a team but don't also evolve the cultural elements that surround that team, it can place an extra burden on people to navigate the "metagame" of exclusion in their work environment *while also* delivering successful solutions for the business.

Ability is one of the few categories that transcends all other types of human diversity.

We are all born and gain abilities as we grow. We lose those abilities as we age. As we move through life, our abilities change as a result of illness or injury. They even change when we move from one environment to the next. Our vision changes when we move from a dark movie theater into the bright sunlight. Our ability to hear a conversation changes from a quiet elevator to a crowded party.

Human ability, in its many physical, cognitive, and societal forms, is a building block of design. A person's capabilities and limitations are always a factor in how successfully they interact with a solution.

Furthermore, something extraordinary happens when we make human ability our first consideration. Everyone can relate to the idea that abilities are limited and ever-changing. It rings true regardless of nationality, professional training, unconscious biases, or worldview. It enables a common ground from which to start designing inclusive solutions.

Of all the biases that designers bring to their work, ability biases are the sneakiest.

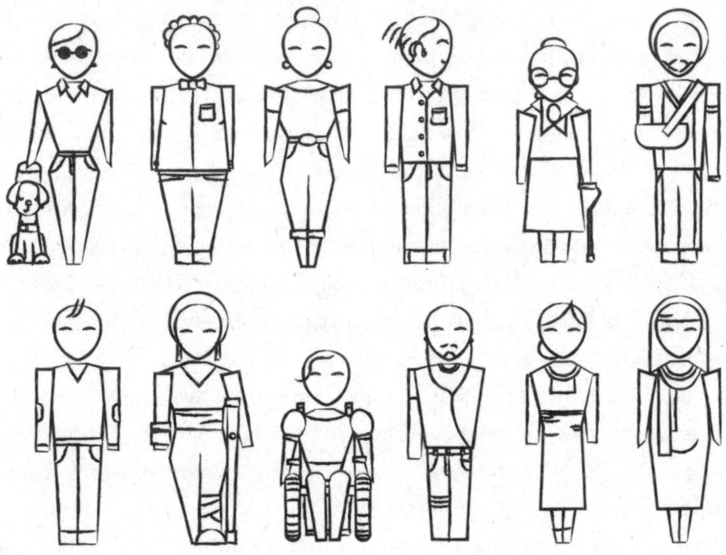

Figure 4.3
Designing with our own abilities as a baseline can lead to solutions that work well for people with similar abilities, but can end up excluding many more people.

An ability bias is a tendency to solve problems while using our own abilities as a baseline. When we do so, our solutions end up working well for people with similar abilities and circumstances, but can exclude a much wider group of people.

Even the most empathetic designer will typically create a solution that she herself can see, hear, and touch. She'll use her own logic and preferred ways of communicating. Her eyesight acuity, hand dexterity, and language fluency will influence the way she creates solutions. Even the design tools that she uses to create a solution will reinforce her ability biases.

Ability biases aren't inherently bad. They aren't necessarily something to eliminate. In fact, they can be strengths. Once a designer develops the skill to recognize their own ability biases, they can start to recognize the ability biases in other people as well.

Porter's ability biases as a gamer are speech commanding, strategic coordination, and navigating problems through experimentation. But he doesn't simply design solutions that work for his own abilities. As an inclusive designer, he creates solutions that consider a range of abilities beyond his own, so that a wider audience can successfully use his designs.

How do we extend beyond our own ability biases? No degree of wearing a blindfold will ever be equivalent to the experience of being blind. The blindfold can actually give designers a false sense of empathy, especially if they attempt to simulate disabilities without ever meeting or working alongside people with disabilities.

If we are designing a solution that will be used by millions of people, and our ability biases are inevitable, where do we start?

To unlock this conundrum, I invite you to consider diversity through the lens of *human interactions*. That is, our interactions with each other and with the world around us.

In 2011 the World Health Organization published their *World Report on Disability*, referring to disability as "a complex phenomenon, reflecting the interaction between features of a person's body and features of the society in which they live."[1] This is also known as the social definition of disability. These points of interaction are where mismatches occur.

For designers, this can open a new mindset. It's a profound shift from thinking about disability as a personal health condition

Disability

personal health condition

Disability

=

mismatched human interactions

Figure 4.4
When we think about disability in terms of mismatched interactions, it highlights the responsibilities of people who make solutions.

outside of the range of "normal." When we consider disability as a mismatched interaction, it underscores the responsibilities of the designer. Every choice we make either increases or decreases the mismatches between people and the world around them.

A well-loved example of mismatched design that led to innovation is OXO tools. Betsy Farber was having difficulty using kitchen utensils due to arthritis in her hands. The thin metal handles of objects, like a potato peeler, were difficult and painful to hold. Her husband, Sam Farber, worked with her to develop a new grip design. The new design was rounded to fill the hand and was made of flexible rubber that could shape to the unique grip of each user. The Farbers created their first set of fifteen OXO Good Grips kitchen tools in 1990.

The design was dramatically more comfortable for Betsy, but also worked well for anyone who had difficulty holding kitchen utensils due to wet or slippery hands.

A DIVERSITY OF WAYS TO PARTICIPATE

Mismatched interactions also arise when we create solutions with only one way to participate. In our gaming example, this can be a controller that requires two thumbs, a high degree of manual dexterity, or a high degree of hand strength. It's also about the range of roles that players can act out in a game (warrior, merchant, coach, athlete, etc.).

Inclusive designers create a multitude of ways for people to participate in and contribute to an experience.

To better understand this, let's take a closer look at several definitions of inclusive design. Although the term has been around for decades, it was largely an academic practice until recently. When I first learned about inclusive design, very few companies were applying it to their work in a repeatable way. At Microsoft, we sought out mentors from universities. Most importantly, Jutta Treviranus and her team at Ontario College of Art and Design. Treviranus founded the Inclusive Design Research Centre in 1993 to focus on ways that digital technology can improve societal inclusion. She has a clear approach to selecting each new cohort of designers:

> We need designers who have experienced barriers. What we want to produce is not a uniform set of individuals with specific competencies, but a group of individuals that can work as a team, that each can contribute a diverse perspective.[2]

Another mentor in our early days was the inclusive design leader that I referenced in the opening of this book, Susan Goltsman. Her definition of inclusive design will always be my favorite:

> Inclusive design doesn't mean you're designing one thing for all people. You're designing a diversity of ways to participate so that everyone has a sense of belonging.[3]

Goltsman led her design projects with what she called the I-N-G's. She would sit and observe how many different human activities were happening in a park. Goltsman would ask, "what I-N-G is most important to this environment?" Maybe it was running, digging, swinging, climbing, or sleeping. Whatever the I-N-G, the next question was always "how many ways can human beings engage in that activity?"

Imagine a playground full of only one kind of swing. A swing that requires you to be a certain height with two arms and two legs. The only people who will come to play are people who match this design, because the design welcomes them and no one else.

And yet there are many different ways you can design an experience of swinging. You can adjust the shape and size of the seat. You can keep a person stationary and swing the environment around them. Participation doesn't require a particular design. But a particular design can prohibit participation.

The same phenomenon applies to technology. If writing stories required a keyboard, computer screen, and fluency in English, the only stories we'd read would be from people who match these requirements. Each feature created by designers determines who can interact with an environment and who is left out.

Building on Treviranus's and Goltsman's guidance, here's a working definition of inclusive design that we developed at Microsoft, after applying it with thousands of engineers, designers, and business leaders.

> Inclusive design: A methodology that enables and draws on the full range of human diversity. Most importantly, this means including and learning from people with a range of perspectives.[4]

We also found it helpful to distinguish inclusive design from related concepts, like accessibility and universal design. Here's a quick primer that guided our work:

Accessibility: 1. The qualities that make an experience open to all. 2. A professional discipline aimed at achieving No. 1.

An important distinction is that accessibility is an attribute, while inclusive design is a method. While practicing inclusive design should make a product more accessible, it's not a process for meeting all accessibility standards. Ideally, accessibility and inclusive design work together to make experiences that are not only compliant with standards, but truly usable and open to all.

Most accessibility criteria grew out of policies and laws that were designed to ensure barrier-free access for specific disability communities. Wheelchair access in architecture only became prominent across the United States after the Americans with Disabilities Act was passed by Congress in 1990. The 1998 Section 508 amendment to the United States Workforce Rehabilitation Act of 1973 mandated that all electronic and information technology be accessible to people with disabilities. The United Nations created a Convention on the Rights of Persons with Disabilities in 2006, as an international agreement to focus on the full societal integration of people with disabilities.[5]

Inclusive design should always start with a solid understanding of accessibility fundamentals. Accessibility criteria are the foundation of integrity for any inclusive solution.

Another concept that is closely related to inclusive design is universal design.

Universal Design: The design of an environment so that it might be accessed and used in the widest possible range of situations without the need for adaptation.[6]

Universal design was born out of the built environment. It is rooted in architecture and environmental design. It emphasizes the end solution, most often one that is physically fixed. The principles of universal design are focused on attributes of the end result, such as "simple and intuitive to use" and "perceptible information."[7]

In contrast, inclusive design was born out of digital technologies in the 1970s and 80s, like captioning for people who are deaf and audio recorded books for blind communities. Inclusive design is now growing into adulthood alongside the Internet.

In some areas of the world, the term inclusive design is used interchangeably with the term universal design. I prefer to make a distinction between them in two ways.

First, universal design is strongest at describing the qualities of a final design. It is exceptionally good at describing the nature of physical objects. Inclusive design, conversely, focuses on *how* a designer arrived at that design. Did their process include the contributions of excluded communities?

The second distinction, initially coined by Treviranus: Universal design is one-size-fits-all. Inclusive design is one-size-fits-one. We'll explore this further in chapter 7 with a technique called a persona spectrum.

Inclusive design might not lead to universal designs. Universal designs might not involve the participation of excluded communities. Accessible solutions aren't always designed to consider human diversity or emotional qualities like beauty or dignity. They simply need to provide access.

Inclusive design, accessibility, and universal design are important for different reasons and have different strengths. Designers should be familiar with all three.[8]

An inclusive designer is someone, arguably anyone, who recognizes and remedies mismatched interactions between people and their world. They seek out the expertise of people who navigate exclusionary designs. The expertise of excluded communities gives insight into a diversity of ways to participate in an experience.

MAKING ACCESSIBILITY ACCESSIBLE

If you aren't already familiar with the basics of accessibility for your field, you're not alone. The good news is you don't need to become an expert in solving everything. You just need to know enough to know when you need to bring in a true expert. You need to know how to recognize accessibility issues and how to design solutions that work well with the assistive technologies that people rely on. Here are four unique challenges that people commonly face when they're new to accessibility.

- *Lack of educational resources.* Accessibility fundamentals are rarely taught in school or by employers. The companies that are leading in inclusion are creating curricula to help engineers and designers learn the basics. There's a community of accessibility specialists who are also producing great educational content. Much of this information is available online in open formats, for free. This will give a general introduction to what you need to know.
- *Complex legal verbiage.* Accessibility focuses on legal standards. The language of these standards can be intimidating, unless you're a lawyer. Some companies hire external agencies to conduct conformance testing for accessibility standards. Accessibility criteria can also be inexact. That is, they might not give specific details on how to create an accessible solution or measure whether it's successful.
- *Finding the signal in the noise.* It can take significant investment and time to build a custom set of criteria that apply to your business or solution. You can also create your own checklist of top exclusion

issues to avoid and test for throughout your design process. In general, we need better ways to validate that inclusive solutions are working as intended in the real world.

- *A highly manual process.* After decades of being underprioritized, the tools for checking accessibility haven't kept pace with advances in technology. As a result, a lot of product testing is conducted line by line, by human beings. This seems to be changing quickly as more people are building the requirements of accessibility right into development tools. Ideally it would be difficult, if not impossible, to produce an inaccessible product.[9]

INTERDEPENDENCE AND COMPLEMENTARY SKILLS

In cultures that overemphasize independence, it's less common to design solutions with *interdependence* in mind. In the United States, there's a deep attachment to the idea of a rugged, lone pioneer (or astronaut, or entrepreneur, or cowboy) venturing out into the great unknown to make their way in the world. When they succeed they're hailed as self-made heroes of strength, ingenuity, and resourcefulness. These stories of independence rarely reflect the truth of our lives, which are full of dependencies.

People with disabilities often depend on human assistants and assistive objects. The skills of these assistants are vital to closing the gaps between the implements of daily life and a person's abilities. And all people are increasingly dependent on technology to engage with the world.

Interdependence is about matching complementary skills and mutual contributions. Thinking back to Porter's example of online gaming, vast worlds are inherently more inclusive when they have a diversity of ways to contribute to the game. These games, like any society, don't thrive solely on the skills of hunters and warriors. The society is a system of interdependent skills, an economy that includes many different types of novices and masters.

When designing a solution, it's easy to treat people as individual entities. Even in the design of shared products, especially in social media, many solutions assume that groups of people behave merely as collections of independent individuals. Each person is treated as a separate entity, blasting bits of information out to other individual people and waiting to count the number of likes that they receive in return.

An inclusive designer thinks in terms of interdependent systems. They study human relationships. They observe the ways that people bring their skills together to complement each other. Inclusive designers seek out a diversity of ways that people build collective accomplishments.

As inclusive designers, interdependence challenges us to think in broader ways about systems of contribution. It prompts us to ask what human activities, which I-N-G's, are most important to the things that we design. Designing for interdependence changes who can contribute to a society, what they contribute, and how they make that contribution.

WE ARE ALL DESIGNERS

The traditional design professions are rapidly changing, especially in areas of technology where the required skills change so quickly that many universities are struggling to maintain a relevant curriculum. Much of today's design work isn't limited to people with the word "design" in their professional titles. Among those evolving design roles there is a new category of skills in inclusive design.

This is critical in many areas of society, but urgently needed in technology design. As technology permeates intimate areas of our lives, design becomes an intimate act. People invite products into their homes. They share their secrets with personal devices. They use digital technology to facilitate moments of sharing, celebration, and mourning. Each of these technologies is built by people who are making assumptions based on their own biases. For better or for worse, these choices can also determine who gets to participate in society.

People who've experienced great degrees of exclusion can translate that expertise to the solutions they create. Their experience can shape every other aspect of the cycle of exclusion. But this strength isn't the result of being considered "different" from other designers by some facet of demographic diversity.

Their expertise stems from being familiar with exclusion and what makes it a universal human experience. This knowledge can make it easier to recognize the exclusion that many more people face, bringing a greater appreciation of what is gained when a design process is truly open to diverse perspectives. This is how inclusive designers learn to flex their ideas and combine their strengths.

...

TAKEAWAYS: BUILDING THE SKILLS TO RECOGNIZE AND RESOLVE MISMATCHES

Exclusion habits

- Creating solutions with only one fixed way to participate.
- Creating solutions using our own abilities as a baseline, known as ability biases.

- Treating accessibility and inclusion as an afterthought or only meeting the minimum legal criteria.

How to shift toward inclusion

- Consider diversity in terms of human interactions and how people change over time.
- Identify ability biases and mismatched interactions that are related to your solution.
- Create a diversity of ways to participate in an experience.
- Design for interdependence and bring complementary skills together.
- Build a basic literacy in accessibility, and grow a depth of expertise in the specific accessibility criteria that are relevant to your solutions.
- Adopt a more flexible definition of a designer. Open up our processes and invite contributions from people with relevant but nontraditional skills.

5 WITH AND FOR

How generations of exclusion are made and broken.

The next time you're at a street corner, waiting to cross, take a moment to notice the details of its design. Note the orientation of buildings, the intensity of sounds or crowds of people. Who designed that street corner? How are their choices influencing your ability to move? Or to interact with other people?

Of the many professionals who contribute to the design of our world, architects have to achieve some of the most stringent criteria before being allowed to practice. The degree of training and testing required to become a licensed architect is incredibly high. After completing the necessary education, the credentialing process requires years of experience and thousands of dollars to complete. The whole process can take 12.5 years, on average.[1]

With public safety at stake, this makes sense. In contrast to the industry of video games, where open-source platforms are making

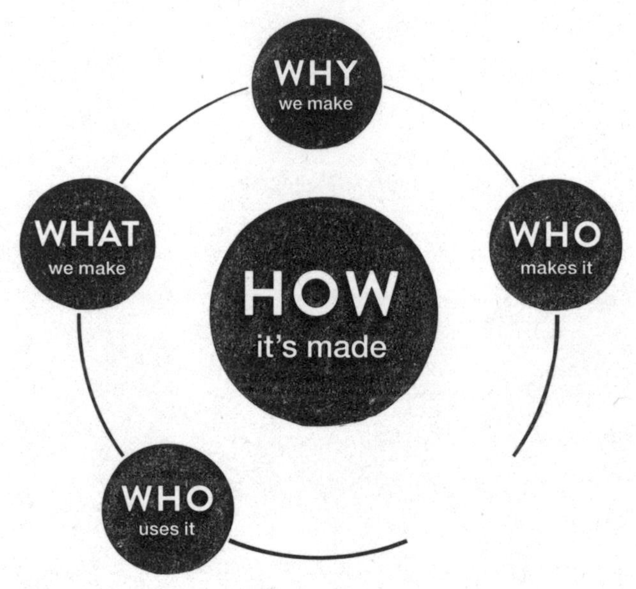

Figure 5.1
Cultural context and preexisting methods can perpetuate the cycle of exclusion.

it easier for anyone to become a game designer, it's hard to imagine the same thing happening in architecture.

This makes the role of an architect a useful lens to examine *how we make*. There's a greater context, beyond the individual designer, that influences exclusion and inclusion.

In addition to business criteria and technical requirements, a design is shaped by the history of events that precede it. This means that shifting a cycle of exclusion toward inclusion isn't simply a matter of designing an object in new ways. We also need to disrupt the momentum of how things have been done for a long, long time.

A CITY OF DESIGN

The whirring of power saws and high-pitched beeping of construction trucks fill the downtown streets of Detroit. The skeletons of new high-rises stand tall alongside historic buildings with some of the most beautiful examples of art deco in the country.

In 2015, Detroit became the first U.S. city to be named a UNESCO City of Design. The distinction is an acknowledgment of the city's rich architectural heritage and a long list of influential local architects and designers who defined American Modern design.[2] While Detroit is well known as the heart of the American automotive industry, its artistic strengths are equally impressive.

In my first visit, Tiffany Brown guided me through the city. An architectural designer and native Detroiter, Brown is exceptionally talented at building connections with people. We wove in and out of parks, restaurants, and cathedral-like lobbies of office buildings. She moved at a breakneck pace, sharing vivid details about each location. The nuances of each design and the names of the architects were all new to me.

Brown describes the features of architecture through the stories of the people who gather there to celebrate, mourn, create, and build. The best restaurants for birthday parties. The murals in a quiet alleyway that transforms into a concert venue on weekends. She has a keen sense of how spaces bring people together and how design creates a sense of belonging.

A growing number of architects, like Brown, believe the people who inhabit a space should contribute directly to its design. Her firsthand experience with how design can fail communities motivated her to pursue architecture. She understands how designing *for*, not *with*, people can lead to exclusion.

We pass through a small triangular park with the words "Paradise Valley" engraved on the ground. Brown points out vacant storefronts that once formed a thriving economic center for hundreds of Black business owners. People gathered in these streets to hear jazz legends like Dizzy Gillespie, Ella Fitzgerald, Duke Ellington, and Louis Armstrong. This park marks the north end of what used to be Black Bottom, named for its rich marsh topsoil, one of the few neighborhoods where African Americans were allowed to live in the 1920s through the 1950s.

In the early 1900s Detroit was a beacon of economic opportunity, especially for African Americans. The city's population exploded from less than 500,000 in 1910 to over 1.5 million people in 1930.[3] Many of these new residents were African American migrants from the south.[4] Neighborhoods like Black Bottom were places whose inhabitants knew and looked out for each other. Residents intermingled with national heroes, like boxer Joe Lewis, and there was a palpable belief in the community that anyone could be great.

Over the course of the decline of the local automotive industry, the nationwide recession in 2008, and the city declaring bankruptcy in 2013, more than one million people left Detroit, likely never to return.

Urban planning and architecture played a role in this decline. Black Bottom was demolished in the early 1960s to make way for the Chrysler Freeway and a new mixed-income neighborhood designed by Ludwig Mies van der Rohe. The thriving center of Detroit's African American community was destroyed and most Black Bottom residents moved to public housing projects across the city.

Brown drives us to the edge of what once was the Brewster-Douglass housing projects. The pace of construction is dramatically softer. The pounding of a hammer starts and stops from inside a brick shell of a partially demolished home. Across the street is a common sight in Detroit: a large lot, empty and overgrown, surrounded by a wire fence with a bright white sign reading "road closed."

The Brewster Homes, which later grew to become Brewster-Douglass, was the first federally funded housing project in the United States. It was opened with a ceremony attended by Eleanor Roosevelt in 1935. The Brewster Homes promised to be family-oriented residences for working-class African Americans who, at the time, had limited housing options. These first homes were largely funded by private benefactors. For several decades, these homes were well managed and maintained by the federal government.

The promise of family-oriented housing started to change when the Frederick Douglass towers were added to the Brewster Homes in 1951. The stark, cramped design of these twenty-story block-shaped towers was a far cry from the original town homes.

Figure 5.2
A Douglass tower standing next to the original Brewster Homes in Detroit.

This style was typical of the rising modernist movement in architecture that borrowed heavily from a vision set by Swiss-French architect Charles-Edouard Jeanneret-Gris, known as Le Corbusier. His concept of "towers in the park" was just one example of his mathematically precise approach to architecture, based on *béton brut*—French for raw concrete.

Le Corbusier's approach to architecture made a strong impression on Robert Moses, a man known as the "master builder" of New York City in the mid-twentieth century. He was also demonstrably racist, as detailed in Robert Caro's Pulitzer Prize-winning biography, *The Power Broker*. Moses built extensive parks, beaches, roadways, tunnels, and bridges that transformed New York City during the decades of his tenure as an appointed public official. And he did so with a keen focus on excluding communities that he deemed unworthy to enjoy them. In one infamous example, he lowered the height of overpasses to prevent the passage of public buses, the primary mode of transit for low-income and African American residents.

One of the practices that Moses embraced, in the footsteps of Le Corbusier, was to demolish what he deemed "blighted" areas, displacing over half a million residents in pursuit of his vision. He would raze entire neighborhoods to the ground, creating a blank slate for new housing projects and expressways. This top-down approach to planning architecture became a beacon of modern urban development, and was replicated in cities across the United States.

Through the 1950s and 1960s Congress allocated fewer and fewer funds to building and maintaining high-quality public housing. As they did, the design quality declined. High-rise towers, often with one elevator to service scores of families, became a design standard for public housing projects across the United States.

With reduced funding, restrictive policies, and neglected maintenance, the buildings of Brewster-Douglass fell into disrepair in the late 20th century. Once home to nearly 10,000 Detroiters, they were marked for demolition, with the promise of rebuilding people's lives on a foundation of entrepreneurship and training for new jobs.

The last residents were moved out in 2008. Federally funded Section 8 housing vouchers were the only way that many people could afford to live outside of the development, which meant moving into private landlord-owned homes in disparate areas of the city. Many residents protested. Some refused to leave their homes and were forcibly removed.

This story isn't unique to Brewster-Douglass. The pattern of destroying Black and low-income neighborhoods to make way for "urban renewal" projects has been repeated across Detroit for decades. As the city's design evolved, it gained new features that excluded people from each other and from economic opportunities.

A racial distribution map underscores this point. Today, the population of Detroit is roughly 700,000 and over 80% are African American.[5] The surrounding metropolitan and suburban areas are predominantly Caucasian. The spaces between communities are marked by clear physical boundaries that correspond to major roadways.

The final stages of the Brewster-Douglass demolition were supervised by Brown in 2014 while she was working for the architecture firm Hamilton Anderson Associates. During the project she met with former residents who arrived to watch as buildings were being deconstructed. They shared family stories, remembered loved ones

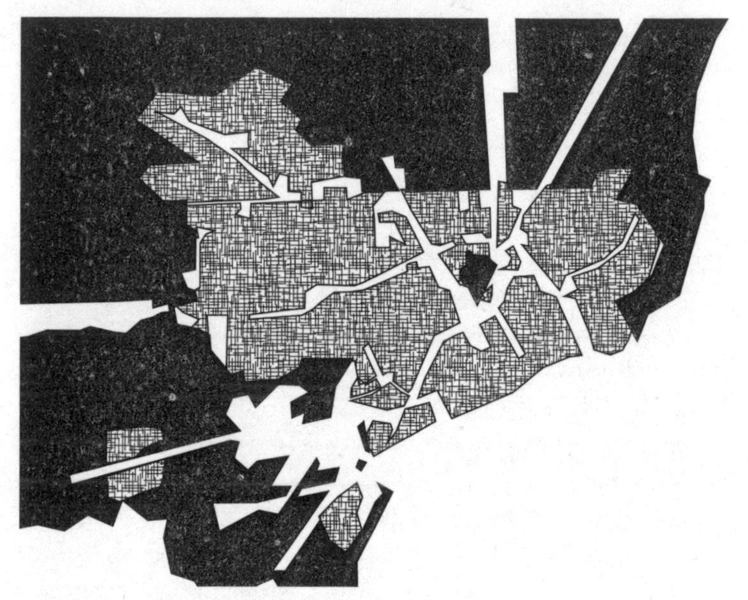

Figure 5.3

Racial distribution in Detroit is sharply segregated along roadways and waterways (shown in white). The population of the areas shown in gray is over 80% African American, and closely follows the boundaries of the city of Detroit. The population of surrounding metropolitan and suburban areas, shown in black, is over 80% Caucasian American. (Based on data from the 2010 U.S. Census.)

lost, and recalled the places where lifelong friendships took root. People gathered to recognize this place that shaped their lives in profound ways.

The Brewster-Douglass demolition held special meaning for Brown because the complex shared a lot of traits with her childhood home, Herman Gardens, several miles away. As we drove along the edge of the now empty multi-acre lot, she pointed out her schools and the home her family was relocated to when Herman Gardens was demolished.

These important spaces from Brown's life now stand vacant and deteriorating. These places influenced her choice to pursue architecture. Not out of passion for the buildings, but for the countless neglected residents that they represent.

Many displaced residents, including Brown's grandmother, returned to live in a small cluster of new homes at the edge of the former Herman Gardens site. From her front porch, we watch her grandchildren climb atop a playground while she recounts the city's promises of new economic development that never materialized.

She can vividly recall the original community and each space where people gathered. Once the homes were destroyed, the local schools lost students and were closed. Businesses lost customers and ceased to exist. She can describe in detail which aspects of Herman Gardens' design worked and which didn't.

Brewster-Douglass was built *for* low-income African American families. But who created the design? Who chose what to tear down and what to rebuild? What was their cultural agenda? Its construction and deconstruction happened without any meaningful inclusion of residents. This practice isn't isolated to architecture. It happens in many areas of design, including digital environments.

When we break from the past, *how* we get to the next stage has an impact on who can move forward with that progress and who's left out. What does it take to shift long-standing cycles of exclusion?

400 FORWARD

From her years as a young resident in public housing to her career as an architectural designer, Brown has encountered many layers of designs that contribute to exclusion. And as she works toward her architecture license, there's one number she always keeps in mind: 400.

In the history of the profession, just over 450 architects have been African American women. In 2017 there were approximately 110,000 licensed architects in the United States. Just over 400 of these active practitioners, roughly 0.3%, are African American women.

The City College of New York conducted a comprehensive study of inclusion in architecture and concluded that the extremely low percentage of African Americans in the profession, just 2% across genders, is the result of specific inequalities in the process of becoming an architect and the support received once new architects are in role. Improving inclusion in the profession, they state, requires deeper study of the "considerations of how minority youth are socialized to think about the field of architecture, and an understanding of the influence of social/peer networks, family guidance, and educational awareness."[6]

The credential process is not the only way to describe the profession. There are countless African American builders and architects who have shaped the American landscape. But credentialing matters, because critical decisions about the future of our cities are often made from positions of seniority and power. And these leadership positions are heavily gated by required credentials.

Brown works with young people as a way to shift the future ranks of leadership. She co-founded the Urban Arts Collective, an organization that uses art, music, and interactive workshops to introduce minority students to architecture and design. Her most recent initiative, 400 Forward, is aimed at empowering the next 400 African American female architects.

Here's how she describes that journey.

How did you choose architecture?

Growing up in Detroit's inner city, we didn't have doctors, lawyers, and engineers running to our schools for career day. We had little access to the arts and zero exposure to architecture. Something like architecture surely felt out of reach for us. I was a student at what was considered a failing public school, and a college recruiter came to an assembly when I was in 12th grade. That fall, I was studying architecture. It changed my life, and also changed my outlook on life.

Returning to my old neighborhood to construct new housing was one of many full circle moments for me. Many of my family members still live there, many of my friends. I've been asked if I would change the way I grew up. My answer is always no. It is a place where I gained friendships that started from kindergarten and last to this day. My parents' generation can say the same thing. To some, it was one of the most dangerous places in the city. To us, it was the only community we knew and called home.

I was once convinced the term "disadvantaged youth" was a proper title for young people deemed unable to achieve success because of their environment or social class. That they were less likely to have, let alone achieve, their goals.

I remember being referred to as a disadvantaged youth by educators, if and when they did come to my school. I did not like how it made me feel. I once believed there were certain things beyond my reach because of my upbringing, because I'm female, and because of my skin color. But then I realized this term did not define me. I began to see it as the reason I would prove them all wrong. I am a "disadvantaged youth" who now has three college degrees, two in architecture.

How does a designer shape inclusion?

I would sum it up with a Latin expression, "Nihil de nobis, sine nobis" or "Nothing about us, without us," which was made prominent by the disability rights movement. This phrase personifies the idea of designing with a community: that no course of action should be decided without total contribution from the people affected by that course of action. Especially groups of people who have been socially excluded, categorized by disability or cultural heritage.

This expression has largely been disregarded in relation to design. As an example, I recently rode the new Q-Line transit system in Detroit. The design of this service emphasizes electronic forms of payment. When it first opened, none of the kiosks offered instructions on how to pay with cash, a secondary option that was only available inside a train. This alienated people in the city who must use change or cash. As a result, it became a way to transport a certain class of people from midtown to downtown and bypass those excluded people. It rides alongside the city bus, stopping at the same traffic lights, held up in the same traffic.

I have personally experienced the outcomes of designing for a particular demographic. The results can be oppressive and unjust. Inhabitants are made to feel powerless and dependent.

What's your approach to inclusion?

While studying architecture, I realized I learn in ways that are different from many of my peers. I did fine in high school, but had a hard time adjusting to college. There was a disconnect between my thought process and how coursework was presented to me. A lot of students with similar upbringings run into this issue and might think that design is not for them.

I need to take the time to get to the root of a problem before I can try solving it. I have to see the big picture and how things flow together. Making sense of how things are related to each other helps me make it personally meaningful.

I also find meaningful connections by meeting with people who are going to be affected by the design.

My passion is to create an entirely new leadership base for architecture so that the profession reflects the communities we serve. Diversity paired with passionate creativity improves the quality of life. It should outweigh this "superhero" approach to design that we have today.

How do you make design accessible?

There was a student who was struggling in one of our architecture workshops. The task was to create a city block, using Lego pieces. She didn't have a logic behind why she'd laid out her city block in a particular way. As designers we try to have some kind of reason for what we did and some kind concept that we start with. Rather than jumping to a design and then rationalizing why we did it. You get to a design because something led you there.

I asked her to think back to something that was emotional for her. She had placed a lot of stores along the front of the city. She explained that shopping was something she did with her mom when she was alive.

I knew right then that she had an emotional story to tell. We talked as she created this block filled with things that she liked to do with her mom. She didn't have to understand design thinking. She just needed to take a step back and clarify the reason why she was making certain choices. Why is this needed? Why is this important?

I want to make architecture and design feel reachable for kids in neighborhoods like the ones I grew up in. This includes them engaging with someone who's from the same type of neighborhood. Someone who has defeated the same odds they face and became successful.

When she describes inclusion, Brown focuses on the greater context. The historical moments that led to an existing design. The orientation between designers and users. The leaders who initiate wide-sweeping changes. Above all, her lens highlights the key to making inclusive solutions: including the expertise of excluded communities.

When people are excluded by mismatched designs, they grow intimately familiar with the nature of the exclusion and how it might be resolved. Given the opportunity, they can apply this expertise toward building inclusive solutions.

The next chapter demonstrates how someone who understands the mismatch can resolve it in creative ways. And how this differs from a person who tries to make a solution based on stereotypes or benevolent pity for someone they perceive as unlike themselves.

...

TAKEAWAYS: HOW GENERATIONS OF EXCLUSION ARE MADE AND BROKEN

Exclusion habits

- Creating mismatched interactions and obstacles to leadership positions that influence design decisions. In particular, barriers that aren't directly correlated to the skills of the profession, such as financial requirements, rigorous scheduling, or only offering one format of test to gain credentials.

How to shift toward inclusion

- Zoom out, consider the system of relationships that led to a solution. Ask how previous designs might be important to future designs.
- Identify a personal connection to an aspect of inclusion that is meaningful to you and your community.
- Study the history of how a solution came to be. Why is it shaped a particular way? Who influenced those decisions, and what was their motivation?

6 MATCHMAKING

How exclusion experts resolve mismatches.

There's a rise in interest in designs that have a positive social impact. A number of projects are focused on "designing for" a community of people that's presumed to be disadvantaged. New technologies for students in developing countries. Design contests to create solutions for elderly people or people with disabilities.

While these are often well-intentioned, there are some potential pitfalls to designing for people with this superhero-victim or benefactor-beneficiary mindset. It can lead to specialized solutions that cater to stereotypes about people.

US AND THEM

To illustrate the problem let's consider the Dodge LaFemme, a car designed specifically for women, brought to market in 1955 and canceled in 1956. The car was pink, inside and out, decorated with

Dodge's LaFemme Is First Automobile With a Gender—It's Female

If Dodge starts a trend with its LaFemme hardtop, license blanks will require owners to state the sex of their cars! Designed for the woman, LaFemme is truly feminine. Its interior is pink throughout. Mounted on the backs of the front seats are two matched pink leather accessory cases. One contains a pink leather handbag with matching lighter, compact, lipstick and other items. The other holds a pink rainwear outfit, including umbrella, that matches the upholstery. The car has an engine, a transmission and a differential, none of which is pink!

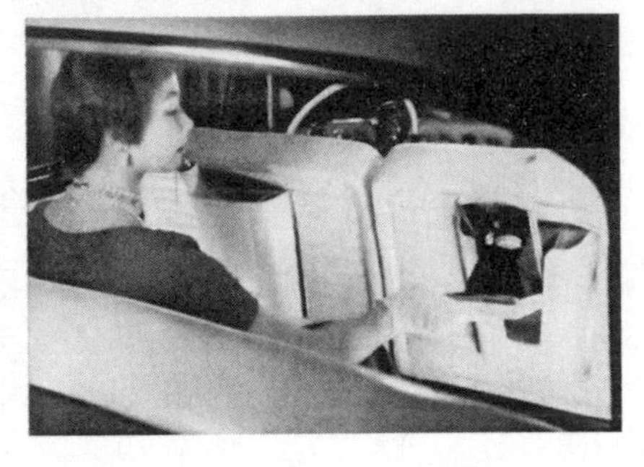

Figure 6.1
A review of Dodge's LaFemme, published by *Popular Mechanics* in July 1955.

small roses. It featured a fully equipped matching purse that fit into the back of the passenger side headrest. It was marketed with the headline "By Special Appointment to Her Majesty . . . the American Woman."

While it's somewhat easy to dismiss this as an artifact of a bygone era of male chauvinism, let's also consider the failed launch of Bic for Her in 2012. This was a line of pens designed specifically for women that were thinner than standard pens and available in pastel shades of pink, purple, and turquoise. It was marketed on Amazon as having an "elegant design—just for her!" and a "thin barrel to fit a woman's hand."

Thin barrel or not, the pens are now a hallmark example of how not to design and market your product to women, thanks to writer Margaret Hartman who sparked thousands of people to write entertainingly sarcastic reviews on Amazon.com. The product was quickly removed from market.

In a more serious example, the automotive industry conducts safety testing with models of humans, also known as crash test dummies. For decades these models were made to match the average male body type, though it was widely known that women were significantly more likely to be injured in a car accident.

In 2011 the federal government started an initiative to reduce demographic disparities in public health. Car accidents ranked high on the list of public health risks. Passenger-side safety ratings plummeted as cars were tested with a petite female crash test model that was 4 feet 11 inches tall and 108 pounds.[1] That year, studies revealed that a female driver, wearing a seatbelt, faced a 47% higher risk of death or serious injury than a male driver.[2]

Decades of design choices where made based on average male-sized testing standards. Engineers and designers were trained to optimize to these standards. It wasn't that the cars were suddenly less safe. They had always been less safe. It just hadn't been recognized as a problem.

This wasn't a sex-specific disparity. The average male crash test model is 5 feet 9 inches tall and 172 pounds. Once the industry started using a range of body types in their safety testing, there was an improvement for any person whose body didn't match the design of the male crash test model, across all genders, sizes, and ages.

We can see from these examples how the deep perceptions people have about one another can be manifested in the design of products and environments. Many features of the most famous American cityscapes were constructed with a specific intention to exclude groups of people from social and economic opportunities. Even after those structures are demolished, retrofitted, or long gone, it's striking how an us-and-them mindset continues to manifest in the designs around us.

Some designs are intentionally created to exclude groups of people. It isn't always a matter of mindlessness or forgetting to consider a community, but a targeted act of discrimination motivated by racism, ableism, sexism, classism, or related motivations to exploit power. Rectifying these forms of exclusionary design requires a shift in culture and a laser-sharp focus on changing the root of that exclusion.

An unspoken hierarchy also appears when attempting to design solutions for groups that are perceived as needing help. Without an authentic and meaningful understanding of a person's life experiences, stereotypes can prevail. All too often, designers and architects perceive the recipients of their solutions as "other people." This mindset distances the designers from people they perceive as disadvantaged beneficiaries of their design.

The problem is separation. It's rooted in the ways we categorize human diversity. The most common ways that we group people by diversity are single dimensions like ability, gender, race, ethnicity, income, sexual orientation, and age. Even if we know that people are more complex than a single dimension, businesses regularly try to solve problems based on these monolithic groupings.

Many of these demographic categories have more to do with business or social power structures than with how people actually interact with the world. It's unclear why a software designer needs to know the gender identity of a customer, or whether or not they have two X chromosomes, in order to create a better way to organize photos. Unchecked assumptions about any group of people, especially when treated as a monolithic group, might misdirect us toward ineffective, even offensive, solutions.

How we categorize people shapes *how we make*. With a growing interest in participatory design methods across many professional fields, how is that participation to be facilitated? What kinds of questions do you lead with? Do you meet people in their homes, or require them to visit your office? What tools do you ask them to use when providing input to the design, and are they comfortable using them? Meaningful inclusion is much more than hosting listening tours, focus groups, or interviewing people on the street.

One way to start is by building an extended community of "exclusion experts" who contribute to your design process. These are people who experience the greatest mismatch when using your solution, or who might be the most negatively affected. Develop meaningful relationships with communities that contribute to a design. Designing with, not for, excluded communities is how we put the *inclusive* in inclusive design.

DISRUPTIVE CHANGE

In the pursuit of innovation, it's common for teams to focus solely on the functional elements of design. It's equally important to understand the emotional considerations of a design, in particular the familiarity people have already developed with an existing solution. What are their patterns of using a solution? What makes these patterns important to their lives?

Consider a person who takes a carefully planned path through a city to make it on time to work every day. Or the ways they organize important files in their personal computing devices. Something as simple as changing the name of a feature in a software application or a street name in a city could be disorienting for them.

This exclusion habit is often motivated by economic factors. Change for the sake of newness. Growth for the sake of progress.

Delight for the sake of differentiation. Fixing perceived disorder into order. Along the way, design changes can disrupt human patterns and relationships. Especially when the problem solver, whether an architect, designer, engineer, or business leader, presumes that their own professional expertise supersedes the life expertise of people who are affected by those changes.

The same thing happens in digital spaces when we change products that people are familiar with. Each time we change a design, by adding new features or moving things around, we require people to learn something new. They have to form new relationships and new patterns of behavior.

The problem is, everyone adapts in their own ways. Not everyone solves problems or learns through the same approaches. But when designers make changes to a product or space, their ability biases can lead the way.

As an example, while at Microsoft, I received a phone call one evening from a product leader who was concerned that there were far fewer women using their product than they expected. He was also concerned by the early solutions that teams were proposing to address this issue.

We took a closer look at the patterns of behavior that were happening in the product. We studied the research of Oregon State University professor Margaret Burnett, who has spent over a decade studying the relationship between gender and software. In her GenderMag project, Dr. Burnett identified a set of facets that consistently lead to differences in how software is used by persons identifying as women or men.[3]

One facet, in particular, stood out to us: how people prefer to learn. Dr. Burnett also refers to this as a person's willingness to tinker with new software. She describes a spectrum that spans between two approaches to learning new technology. On one end is a preference to learn through a guided approach, or with the assistance of a human being. On the other end is a high willingness to explore a software interface through trial and error.

The research showed that people who identified as women distributed relatively evenly across this spectrum. There was a wide range of learning styles that different women used when learning new software. People who identified as men, however, clustered heavily toward the end of the spectrum for tinkering and troubleshooting solutions.

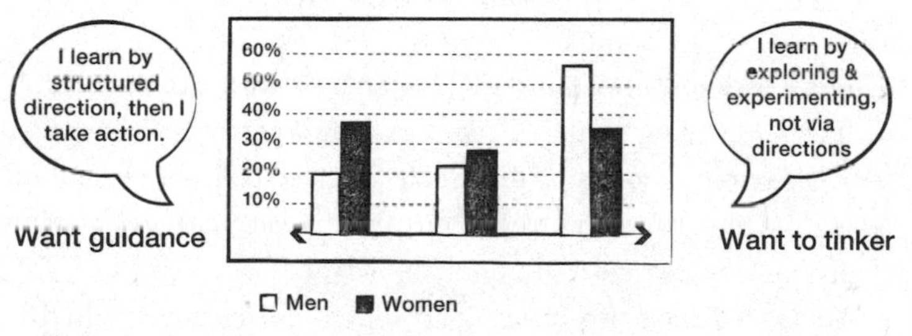

Figure 6.2

A representation of gender distribution as people discover and learn to use software. (Based on data collected by the GenderMag project in 2016. For more information visit www.mismatch.design.)

This insight helped us reframe the problem. Was it possible that our product favored a particular learning style? We restructured our research to recruit people by learning style, and interviewed people from multiple genders, including transgender participants.[4]

We found that people who preferred a guided learning approach, regardless of gender, felt alienated and confused by recent changes we made in the product. They were concerned that important programs had disappeared. Or they couldn't figure out how to complete tasks they'd known how to do for years, because the interface had changed.

It turned out that when we updated our product, we required people to learn something new. But we did so in a way that reflected our own internal learning styles. This differed from how many of our customers learned. After all, what percentage of the general population is trained to think like engineers? Design decisions were made with the assumption that people would just hunt around and try things until they found what they needed, reflecting our team's own learning styles and disproportionately benefitting men.

A SENSE OF BELONGING

This ties back to Brown's insight about her own ways of learning. It isn't about needing help. It's about making a personal connection. Guided learning is one way for people to understand why a solution works the way it does, which helps them feel more confident in their abilities to utilize a solution.

We applied inclusive design and sought out customers, across genders, who preferred a more guided learning approach. The process was similar to seeking out people who have ability biases that complement the dominant ability biases of our teams.

These conversations helped us understand elegant ways to connect customers to existing product guidance, and present it in ways that were relevant to the task they were trying to complete. Even more importantly, the engineering and design teams were united around the idea that we could tangibly include customers who could fill the gaps in our own knowledge. We isolated what we didn't know and sought out this expertise from customers who felt alienated. Then we asked thoughtful questions, and let their answers change our design process.

Disruptive changes can especially be an issue for people with disabilities who might depend on a particular technology to complete essential tasks in their lives. If a software program or website is updated, but the right steps aren't taken to ensure the updates are compatible with assistive tools like screen readers, the resulting changes can literally prohibit someone from doing their job. Or prevent them from using a form of transit that they need to get to work on time. A change to a payment system can impact a person's livelihood.

These learning biases can be further reinforced by feedback from customers. Many companies depend on their most engaged users, people who love products as much as the people who make them, to spend time providing feedback.

How feedback is collected often reflects the preferences of the team that builds the product. As you might imagine, if you only take online feedback, or only provide customer support in English, the only people you'll hear from are people who match this profile. This has a profound impact on which feedback makes its way to the design team.

These feedback channels can also be a signal to customers that they either do or don't belong with the product. With technology, many customers have a tendency to blame themselves for not being able to figure out the changes on their own. Common indicators are customers who say "I feel like technology is moving so much faster than I am" or "I'm probably not smart enough to figure this out." In essence, they feel excluded. The impact on people can be deeply emotional.

Shifting that sense of exclusion requires careful attention to who's missing from a solution and from feedback channels. Whose voices are the loudest and whose are missing? Seek out who's missing and learn about their existing patterns of behavior. Design solutions that bring them successfully through your changes. Provide a diversity of ways to get guidance on what's new and help customers get reoriented with a product they've known and used for years.

We can also shift the sense of belonging by opening up the ways that people can contribute to the design process itself. Contributing to the design of a product or environment, even in the smallest ways, increases the emotional connection between a person and that solution.

The push to accelerate growth and change, for cities or software, is often necessary. But *how* it is implemented is vitally important.

People create emotional connections to a design that make a place, or a product, feel like their own. Introducing change isn't just about breaking apart concrete or bits of code. It's breaking apart human relationships. The result may be that people will leave and never return.

Making a change without disrupting a sense of belonging can be difficult. It's a challenge because it's an emotional choice, not just a rational one. Those emotional considerations are best described by people who are the most excluded from your solution. Or those who stand to lose the most during times of change, including kids who will interact with the next generation of designs. Their contributions will be one of your greatest resources in designing where to go next.

MY HOUSE, MY RULES

Leaders can be powerful advocates for creating inclusion. They set the house rules. Although leadership can come from anyone in a team, there is a unique responsibility for people at the most senior levels of an organization. They must be willing to do the personal work of understanding inclusive design. Certainly there are functional investments that are important, but a senior leader can make or break a culture of inclusion.

When someone in a leadership position declares they are committed to inclusion, some people will be inspired to follow. Others will be skeptical after years of knowing the opposite to be true. And absolutely everyone will be waiting for what happens next.

People are often caught off guard after they declare a commitment to inclusion. The first thing they need to contend with is everything that isn't inclusive. It will show up and make itself known. How a leader listens to, learns from, and invests in shifting these exclusions will be the greatest indicator of their true intentions. Here are four considerations for leaders who want to improve the inclusion of their team.

- *Make promises that you can keep.* Acknowledge the current state of inclusion in your organization and address fundamental issues of access before moving on to other areas of inclusion. Greater damage to inclusion comes from declaring it a promise while having no plan for how to implement the change. Or building new innovations on top of systems that lack basic accessibility. A broken promise is more detrimental than making no promise in the first place.
- *Set an expectation that inclusion is a long game.* Balance the cultural history that led us to where we are today with the reasons why inclusion matters to the future of an organization. Have measured plans for how to address entrenched exclusion habits. These have to account for the tradeoffs in resources that need to be made in order

to build inclusive solutions. Specific people need to be accountable for completing the work. The work is hard and the road is long. But, as progress begins to happen, inclusion can be one of the strongest ways to mobilize people around a shared purpose. The work is meaningful, not just because it benefits overlooked communities, but because it can drive new ideas for growth with untapped ingenuity and fresh thinking.

- *Create a system of incentives and rewards that will motivate people to make inclusive designs.* If the incentive structure for an engineer, designer, or marketer specifically calls out rewards for making inclusion a priority at the beginning of the design process, it demonstrates that an organization is truly committed to inclusion. If investments in inclusion, like accessibility, are treated as an added tax, paying that tax will always be deferred in favor of other business priorities. What we measure shows what we value.
- *Bring people along in the process.* Create a diversity of ways for people to contribute to changing your perspective as a leader. Apply the power of your leadership position to uplifting the excluded communities that are affected by your choices.

It isn't organic. It doesn't happen purely through goodwill. It takes intention, planning, and stamina.

CONNECTING PEOPLE

Inclusive solutions connect people, whether they're in a neighborhood or somewhere on the Internet. The solutions that we build can be economic catalysts for excluded communities. Cities and technologies can bridge people to better opportunities through access to work, education, and social resources.

But how we go about making those solutions sets the stage for who benefits from the opportunities. This is what distinguishes inclusive growth from growth that only benefits a few.

Design influences how people view themselves and their community. The outcomes of design carry the imprints of the professionals who crafted them. They are scratched and scored by the remnants of their creators' thought processes and assumptions. Even after the architect or designer is long gone, their mark endures. The generations of people who live with those designs every day can tell you the exact ways in which that design was a success and a failure.

In times of growth and revival, how we go about creating our solutions can be the difference between perpetuating or transforming the cycle of exclusion. Moments of change are the ideal time to focus on inclusion. The key to success is matching great design challenges with great guidance from exclusion experts.

...

TAKEAWAYS: HOW EXCLUSION EXPERTS RESOLVE MISMATCHES

Exclusion habits

- A "for others" or "superhero" mindset, where pity and stereotypes influence design decisions without any meaningful contribution from excluded communities.
- A top-down approach to making decisions. Presuming that professional expertise supersedes life experience.
- Disregarding existing patterns of familiarity in pursuit of growth. Change for the sake of change.

How to shift toward inclusion

- Identify exclusion experts. These are people who stand to lose the most or face the greatest mismatches with any changes you make to a solution.
- Design with, not for. Facilitate meaningful ways for exclusion experts to contribute to your design process.
- Understand the emotional value that people have already invested in an existing solution. Incorporate these emotional considerations as you create new designs.
- Maintain an ongoing community of exclusion experts who can fill your own gaps in perspective. Build relationships with local organizations that support excluded communities. Understand the role your product plays, or could play, in people's lives.
- Review the techniques you use to collect, sort, and analyze customer feedback. Analyze how the design of that system determines who is willing and able to contribute feedback. Whose voice is loudest? Whose is missing?

7 THERE'S NO SUCH THING AS NORMAL

Testing our assumptions about human beings.

Always remember that you are absolutely unique. Just like everyone else.
—Dr. Margaret Mead, pioneer in cultural anthropology

How many different kinds of human beings are there in the world?

If we aim to create solutions that benefit millions, even billions of people, our solutions need to work across a wide range of human diversity. But that's easier said than done. People can be highly unpredictable. How do we design for so much complexity?

A common exclusion habit is to oversimplify *who uses* or receives a solution. And then we forget to add human diversity back into our design process.

Designers use many techniques to envision masses of people. Many of them are plagued by one dangerous idea: the "normal" human.

Our notions of normal were heavily influenced by a 19th-century Belgian astronomer and mathematician, Adolphe Quetelet. Here's

Figure 7.1
Designers and engineers often use their personal assumptions about people to simplify the ways they think about users.

a quick peek into that story, derived from Todd Rose's exceptional book, *The End of Average*.[1]

Quetelet had ambitions to be as well-known as Isaac Newton. Newton's laws of motion and thermodynamics had brought certainty to seemingly unpredictable phenomena in the universe. Probability and mathematical prediction grew in popularity, sparking new fields of science.

Quetelet turned his ambition toward a new pursuit: using mathematical methods to make sense of uncertainty in human society.

He started measuring human beings and amassing that data into statistical models.

There was one mathematical model that dominated Quetelet's approach: the Gaussian distribution. It's more commonly known as the bell curve. Carl Friedrich Gauss, a German mathematician, developed the proof of a concept that had been introduced decades earlier by Abraham de Moivre, a French mathematician and contemporary of Newton's.

Gauss proved that the probability of an event (like the positioning of an astronomical object, or the flipping of a coin) could be drawn as a normal curve and that the average, or vertical midline, of that curve gave the closest representation of the true nature of that event. The bell-like shape of a normal curve makes it easy to recognize.

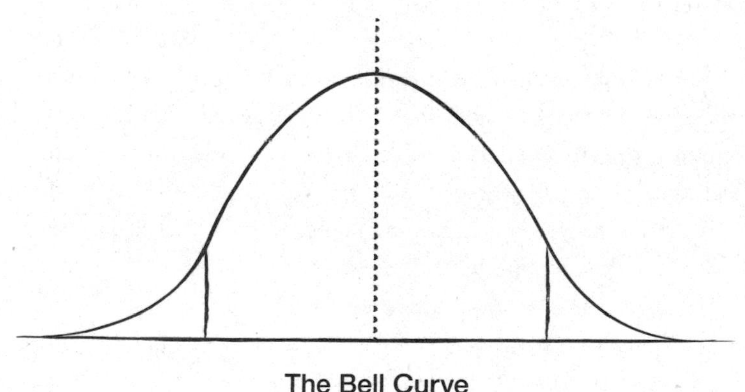

The Bell Curve

Figure 7.2

Normal distribution has had many names, such as the laws of error and the bell curve. The word "normal" was first used to describe mathematical elements of this curve that are perpendicular to each other. It did not imply "common" or "usual." (E. T. Jaynes, *Probability Theory: The Logic of Science* [Cambridge: Cambridge University Press, 2003], ch. 7.)

Quetelet gathered data on human bodies, such as height-to-weight ratios and rates of growth, across thousands of people in Belgium and surrounding areas in Europe. He plotted that data and was astonished to find they mapped to bell curves.

Invigorated by his discovery, Quetelet started measuring many more aspects of human beings, creating physical, mental, behavioral, and moral categories of people. Everywhere he looked, he found bell curves. He became consumed with what he deemed the human ideal, the perfect average measurement across all those dimensions. Quetelet described the perfect face, height, intelligence, moral character, and beyond:

> If the average man were completely determined, we might, as I have already observed, consider him a type of perfection; and everything differing from his proportions or condition, would constitute deformity and disease.[2]

When he published his *Treatise on Man* it was a revolutionary work. In its pages he held that individual people should be measured against that perfect average. From this comparison, one could calculate the innate degree of "abnormality" of an individual person. Diversity and variations in human beings were treated as degrees of error from perfection.

The idea was contagious and enduring.

The bell curve was used to revolutionize existing fields of study and sparked entirely new ones, especially in the social sciences. Normal-based methods of diagnosing illness led to advancements in public health. Quetelet's body mass index (BMI) is still used in many areas of the world to advise individuals on their degree of obesity or fitness. Eugenics, and its horrific assertions about the superiority of select abilities, races, and classes

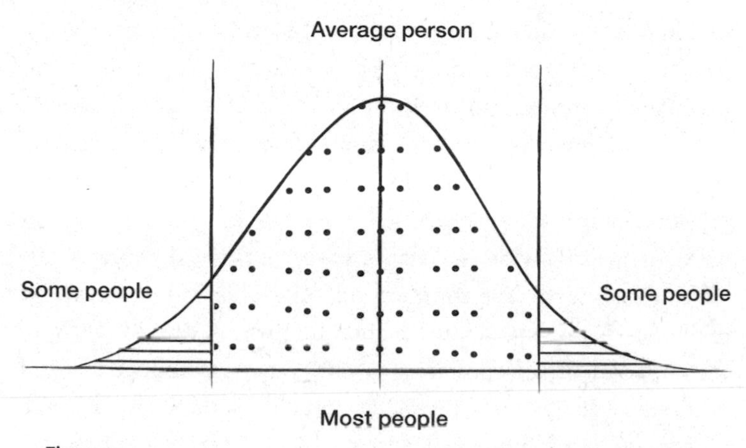

Figure 7.3
When a bell curve reflects a distribution of human beings, it incentivizes designers to target a mythical "average" human in an effort to reach the greatest number of people.

of people, were born out of an idolization of Quetelet's perfect average human.

The conclusions that were derived from Quetelet's approach were used to strengthen the power of some human beings and to dehumanize others.

The power of the bell curve still echoes through the design of our society, from classrooms to computers. Left-handed students are seated in desks made with the assumption that normal human beings are right-handed. Important features of smartphone applications are placed based on where the average user, also presumably right-handed, is likely to reach for them.

A common shorthand for this mindset is the "80/20 rule," originally created by Italian economist Vilfredo Pareto, who demonstrated that 80% of the land in Italy was owned by 20% of the population. Joseph Juran, a pioneer in the field of quality management, translated Pareto's insight into a rule for quality control, stating that 80% of problems come from 20% of causes. In essence, he was separating the "vital few" key sources to a problem from the "useful many" potential sources to a problem.[3]

Over time, many design teams have conflated the 80/20 rule with the bell curve. The common misconception is that the center of the curve represents an 80% majority of the population *and* 80% of the important product problems to solve. The presumption is that if we design a solution that fits the largest bulk of the curve, the middle average, our solution will work well for the majority of people.

This leads many teams to treat the remaining 20% as outliers or "edge cases," a category of work that's often deferred or neglected.

In fact, edge cases can be a useful starting point for creating better solutions. Many exclusion experts use solutions in ways that resemble edge cases. But an edge case implies the existence of a normal, average human. When it comes to design, what if this average human is simply a myth?

EVERYONE AND NO ONE

Rose provides a vivid example from the history of the United States Air Force (USAF).[4] In the 1940s, the first fighter jets were designed to fit the average pilot. The USAF measured hundreds of bodily dimensions across thousands of pilots and used the averages of that data to design the instruments of the flight deck, or cockpit.

Every feature of that flight deck was fixed in place, without adjustability. The assumption was that any individual pilot could adjust himself to overcome the gap in reaching any element of the flight deck that wasn't a perfect fit for him.

However, the Air Force was experiencing a high rate of crashes that couldn't be attributed to mechanical failure or pilot error. A lieutenant and researcher, Gilbert Daniels, studied just ten of those human dimensions that were used in the design of the original flight deck. He measured four thousand pilots to confirm how many of them fit all ten dimensions.

The answer was zero.

Not a single pilot matched all ten dimensions. Everyone differed in at least one way. In essence the USAF had designed a flight deck for everyone, and no one.

This led to the development of new design principles based on individual fit. Innovations like adjustable seat belts, seat heights, and positioning of controls. Once the instruments of a flight deck could be modified by an individual pilot, their safety and performance improved. The adjustments also drastically increased the number, and diversity, of people who qualified to be fighter pilots.

These advancements have influenced the design of many more industrial products. Every time you ride in a car and move your seat, seatbelt, or mirrors into a position that works best for you, you're benefitting from these original innovations in individual fit.

The fighter jet example underscores how designing for the average serves no one, whether the design is for the shape of a desk or a curriculum that emphasizes specific ways of learning. Many designs make people feel like those first pilots, adapting themselves in extreme ways in order to use a mismatched design.

The idea of a normal human being was constructed out of the ambitions of a 19th-century mathematician. It was used to make meaningful advancements in some areas of society, and used to great detriment in other areas. Most importantly, what if the idea simply wasn't true?

All people are variable over the course of their lives. What if our minds and bodies are simply unpredictable? Which human, exactly, should be at the center of human-centered design?

HUMAN-LED, BEYOND HUMAN-CENTERED DESIGN

Personas are a common tool that designers and marketers employ when thinking about who will use their product. A persona is a description of a mythical person, backed by large amounts of research data. The persona might join a name like "Janet" with an image of a woman on a poster next to a description that states, "I'm a soccer mom and freelance consultant who coordinates my family's calendar and has limited time to learn about new technology." Or the smiling face of a man sitting at a computer next to the words "I'm Jim. I'm the lone information technology professional at a startup and I always want to keep up with the latest gadgets."

Like bell curves, personas aim to minimize the uncertainty that's inherent in trying to understand a large group of people. Personas were created to remind designers and engineers that they're building solutions for someone other than themselves. While well-intentioned, personas oversimplify people, without clear direction on how or when to add human diversity back into the design process.

Designers are also taught that the long tails of a bell curve are the edge cases of how people will use their solutions. The exceptional uses. Issues related to accessibility, in particular, are treated as part of that long tail. The bell-curved mindset mistakenly leads us to believe that the market opportunity for accessibility is minimal. Accessibility is treated as an extreme of otherwise "normal" design. In chapter 8 we will review examples that disprove this assumption.

If there is no normal user, there is also no extreme user. There is no such thing as people on the far reaches of the curve. There is no abnormal scenario. There is no edge case. Rather, we need new tools to represent human diversity and challenge entrenched habits of designing for the average.

When personal computers first rose to popularity, it made sense to design graphic interfaces for those computers based on wide-sweeping generalizations about people. Typically, they imagined one person, using one computer, in one environment, completing one or two tasks at a time.

In these early days of digital technology, the variability from one user to the next was also relatively low. Most people were novices at using computers to write an essay or create a spreadsheet. Therefore, it was somewhat easy to make general assumptions about who was using the computer, how they would be interacting with it, and their level of skill. Likely they would be seated in front of a screen and typing on a keyboard in a relatively quiet environment.

Using these assumptions, it made sense for computers to be visually and cognitively demanding, with arrays of icons on a desktop requiring people to hunt for tiny controls hidden in obscure locations.

Today, however, unpredictability in how people use technology is higher than ever. Interactions with technology have undergone an explosion of diversity. Nearly every industry is rethinking their business, products, and services through the lens of digital transformation. It's a necessary consideration for many businesses that hope to remain competitive in the 21st century.

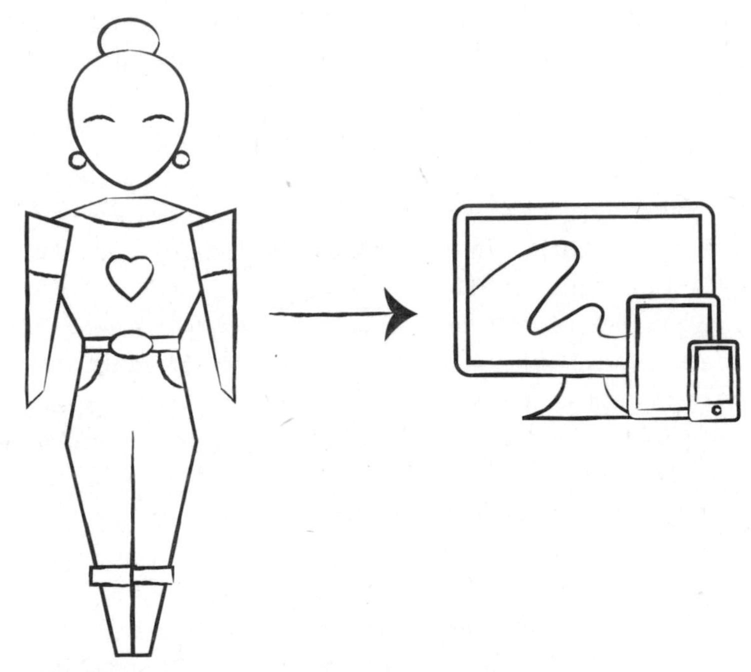

Figure 7.4
Early generations of personal computer interface design were based on a low degree of variability in how people used computers.

One person might have dozens of interactions with digital interfaces in the course of one day, and not just in using a smartphone. People interact with digital interfaces when they buy groceries, follow directions in a car, take a train, buy a cup of coffee, apply for a job, learn a lesson in school, or check out a book at the library.

They use computers in movie theaters, sun-filled parks, noisy cafés, in the rain, and to set bedtime music for their babies. Sometimes they want absolute silence so they can focus. Sometimes

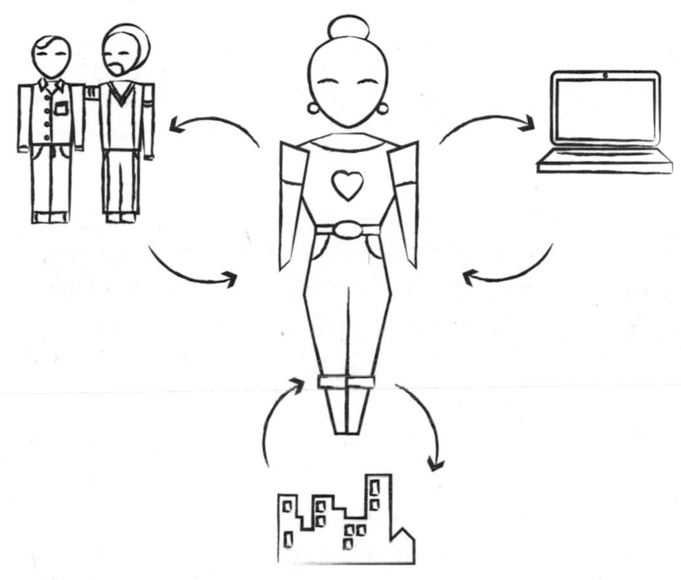

Figure 7.5
There is a dynamic diversity of interactions between people and society, especially as digital experiences facilitate our interactions with more aspects of society.

they want to be interrupted with an urgent notification from a loved one.

This means technology needs to be more than smart. It has to be appropriate to a given person for a specific time, place, and purpose.

It can be overwhelming to even think about such diversity. How can we design for so much uncertainty as people move between environments and devices?

We won't get there on math alone. It's indisputable that mathematical models are a cornerstone of technology design. It's incredibly useful to determine patterns across large groups of people. But when is it appropriate to apply these models to individual human beings? Math has its limits. Describing human individuality is one of those limits. If our aim is to create solutions that are beneficial in people's lives, there are potentially harmful consequences to trying to understand humans solely through math.

Technology has a lot to learn from humans. In particular, from humans who have a deep understanding of mismatched interactions. Their expertise guides better questions, as well as better solutions.

LEARNING FROM HUMAN EXPERTISE

Dr. Margaret Mead is often credited with transforming how we study human cultures. She brought attention to the role that culture plays in shaping individual personality. She helped popularize cultural anthropology and at the time of her death was one of the most famous anthropologists in the world. She called out inherent biases in intelligence tests at a time when many sociologists were churning out studies in attempts to prove racial and gender superiority.

While Quetelet-like methods inspired many academics to pursue bell-curved conclusions about humans, Mead conducted deep studies of cultures that were vastly different from her own. She spent extensive periods of time living with and observing the native cultures of Oceania, especially in Papua New Guinea.

An excerpt from the *New York Times* 1978 obituary of Dr. Mead reads:

> Her conclusions were based on detailed observation, and if she did not conduct anthropometric tests or produce statistical surveys she did convey her subjects graphically.

> Dr. Mead settled down with the people she was studying. She ate their wild boar, wild pigeon and dried fish; helped to care for ill children, and gained the confidence of her informants. At one time she built a wall-less house so she could observe everything around her. She possessed a trait unusual in anthropologists of her time, an ability to shed her Western preconceptions.[5]

Dr. Mead focused on participation as a way to reveal patterns and conclusions. As she learned, the people she studied were in the lead. They were the experts in being human.

When Quetelet gathered data on humans, he was consumed with quantifying the face, body, and full character of a singular ideal average man. His motivation was to fit all aspects of society to this average while dismissing any human who deviated from that norm as abnormal. This motivation has endured, whether implicit or explicit, through generations of his intellectual descendants, including many current designers, engineers, and marketers.

This matters more than ever as data science grows in prominence. All of us are feeding that growth by interacting with computers that record more information about our behaviors than ever before. But data is just data. It doesn't provide answers. The art of drawing conclusions from data is knowing how to collect, organize, and make sense of it.

Taking time to qualify human depth and complexity, even with a relatively small number of people, can help balance the shortcomings of "big data." "Thick data" is gathered information that explains human behavior and the context of that behavior. It was first described as "thick description" by anthropologist Clifford Geertz in his book *The Interpretation of Cultures*. Thick data is a way of understanding how people feel, think, and react and their underlying motivations.

Together, "big data" and "thick data" can give specific and measurable practices for reintroducing human diversity into the design process. For design, big data is like a heat map that can point out interesting areas to investigate. It sheds light on where patterns exist, especially when people use a solution in ways that don't match a designer's expectations.

In complementary ways, thick data can help us zoom in and understand what's really going on. It helps reveal the underlying reasons why big data patterns exist. It helps us identify the human behaviors that contribute to large trends. Big and thick data can work together to help us understand areas of exclusion in products and environments.

It's challenging, if harmful, to pursue one universal model to guide how we think about all humans. But it is possible to understand in great depth a particular *problem* worth solving for one person.

THE PERSONA SPECTRUM

Earlier we examined the distinction between inclusive and universal design. The former emphasizes one-size-fits-one solutions, the latter emphasizes one-size-fits-all. The persona spectrum is an inclusive design method that solves for one person and then extends to many.

Consider the rise of television, a visual and audio medium.

Captioning was initially created in the early 1970s by the National Bureau of Standards and ABC to make television content accessible to Deaf and hard-of-hearing communities. In 1972, Julia Child's *The French Chef* aired on PBS as the first television episode with captioning.

Today, roughly 360 million people in the world are deaf or have profound hearing loss. Thirty-two million of them are children.[6] Nearly everyone, given a long enough life span, loses their hearing as they age.

Captioning and subtitles are essential tools for making information accessible to millions of people. And many more people benefit from these solutions. In noisy airports or crowded pubs, patrons rely on captioning to access sports and news. Captions are increasingly common in social media when we can't turn up the volume on our smartphones. They also assist new readers or people who are learning a new language.

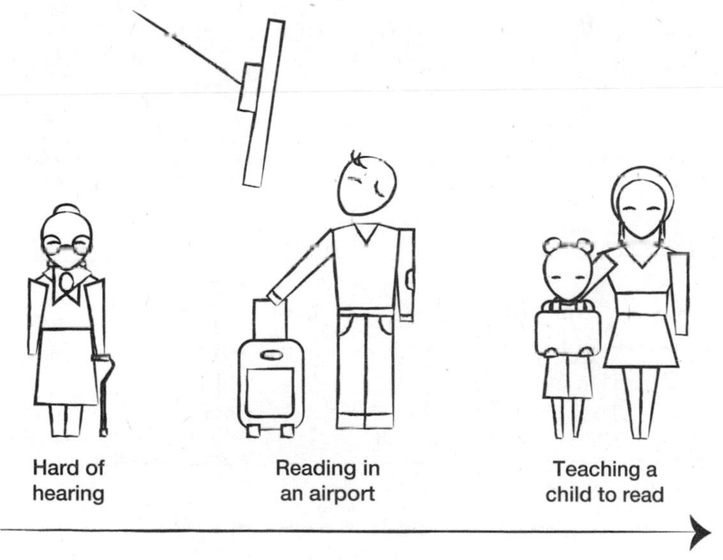

Figure 7.6
Closed captioning, originally designed for Deaf communities, is beneficial to many more people across a variety of environments and circumstances.

This solution initially addressed a mismatch between television audio and the Deaf community. It went on to benefit many more people with hearing loss because of age, injury, or environment.

The persona spectrum can help us create inclusive designs in repeatable ways. Here are four examples of persona spectra based on some physical human abilities. Similar spectra could be drawn for any dimension of human physical, cognitive, emotional, and societal abilities. It all depends on the problem you're trying to solve and the ways that people might interact with that solution.

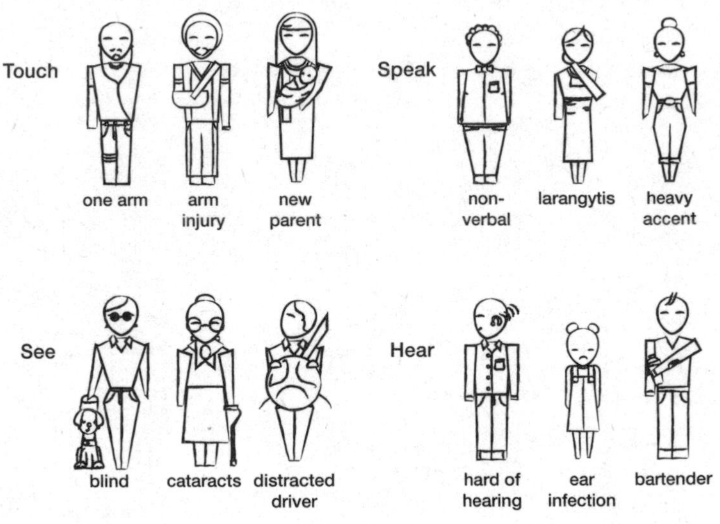

Figure 7.7
There's a spectrum of permanent, temporary, and situational mismatches that people experience based on their abilities and disabilities.

The persona spectrum starts on the left, with the person who experiences the most mismatched interaction. For example, if you were designing a charging station for an electric vehicle, you would certainly need to consider how it would work for any driver who was born with one arm. There would be a lot to learn from talking with people about how well existing gasoline fuel pumps work.

But you could also learn a lot by spending time with someone who's blind to understand how they navigate existing payment kiosks for buying groceries or bus tickets. This perspective is also important for a time when self-driving electric cars become mainstream and will need to work well for people who are blind.

A persona spectrum is more than just a continuum of ability. It's about understanding why people across that spectrum want to access that solution.

An example of this important nuance is Hearing AI, a project initiated by a Microsoft developer, Swetha Machanavajhala. One day, her neighbor showed up at her front door, angry about the loud sound coming from Machanavajhala's apartment. It was her carbon monoxide detector. Machanavajhala, who has profound hearing loss, hadn't heard it.

The experience inspired her to build an application that could alert the user to sounds in their environment. Taking an inclusive design process, her team learned from a range of people who were deaf and hard of hearing. They learned about the importance of Deaf culture and participated in American Sign Language classes and events.[7]

An important theme emerged: no one was interested in a solution that attempted to "fix" or "replace" an ability to hear. The functionality of telling people what a sound was, like a crying baby or a honking horn, felt like an inappropriate attempt to fix a person's deafness.

Rather, people were interested in the emotional information that would guide them to make their own conclusions about the nature of a sound. For example, knowing whether the inflection in a person's voice meant that they were being sarcastic or being angry. As a result, Swetha and team built a solution that visualized the intensity of a sound and the direction it was coming from, so that a user could turn in that direction and determine what it was for themselves.

Tying back to the persona spectrum, it's important to understand what motivates a person to use a solution. Rarely is it a purely functional reason, like knowing the identity of a sound. More often they are motivated by some universal human need, like an increased sense of independence or creating an emotional connection.

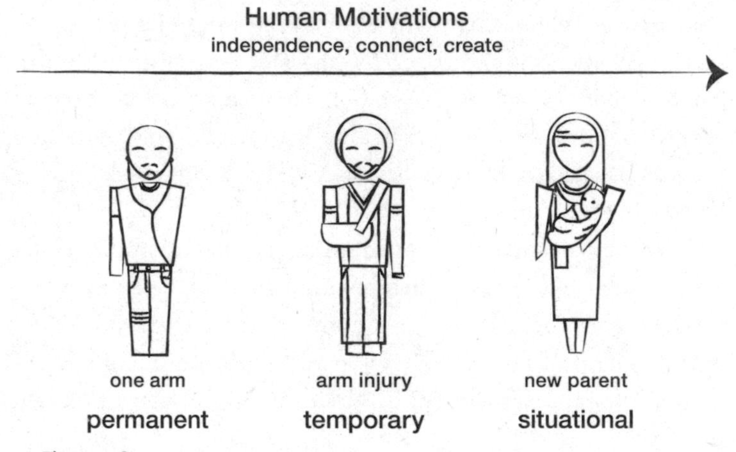

Figure 7.8
A shared human motivation guides how to extend the benefits of a design across a spectrum of people who are excluded on a temporary or situational basis.

These motivations are the glue that connects people across a persona spectrum. When we design for one person who experiences mismatches in using a solution, we can then extend the benefits of that design to more people by asking who else might want to participate but is excluded on a temporary or situational basis.

A persona spectrum can improve the inclusion of existing solutions, extend seemingly niche solutions to broader populations of people, and futureproof your design.

COGNITIVE, SENSORY, AND SOCIETAL MISMATCHES

These introductory persona spectra focus heavily on physical mismatches. Exclusive designs are equally prevalent in cognitive, sensory, and societal areas. Sometimes they can be more challenging to recognize: while a physical barrier often literally stands in your way, these other types of mismatches might be subtler. Here are a few thoughts on how we can extend the fundamentals of inclusive design beyond physical abilities.

With digital experiences, cognitive mismatches are especially important to understand. The human mind is being influenced in so many new ways by applications that constantly compete for our attention. Or augmented reality products that change how people perceive a combination of physical and virtual environments. Building a solution to fit how a person learns, remembers, focuses, uses language, or experiences sensations can be a complex design challenge.

The diversity of the human brain and body is far from being fully understood. Yet there are a growing number of examples of cognitively inclusive products, many of them focused on education and learning environments. Some of these are built with the participation of people with cognitive and sensory disabilities. Others are focused on types of cognitive diversity that aren't disabilities.

Societally, technology is a gateway to accessing economic and social opportunities, especially as people around the world join the

Internet for the first time. There are over 730 million people on the Internet in China, roughly twice the population of the United States.[8] Ninety-five percent of Internet users in China access it using mobile devices.[9]

Many people are joining the Internet through voice-based computers, by talking to their devices as a primary mode of interaction. Augmented and virtual reality are quickly becoming primary ways to access the Internet for millions of people in China.

Global shifts in digital participation lead to interesting design challenges. In areas of the world where touchscreens dominate, most designers are focused on creating graphic user interfaces. These can be seen and touched. In areas of the world where voice-activated interfaces will dominate, designers are focused on speech commanding and conversational design.

Design for societal inclusion needs to consider the greater cultural context that surrounds a design. Complex challenges like language, politics, currency, Internet bandwidth, and social nuances are important factors in designing inclusive solutions.

At minimum, inclusion will depend on how well designers can work beyond the types of abilities and technologies that dominate their immediate environment.

DESIGN FOR ONE, DESIGN FOR 7 BILLION

The more I learn about inclusive design, the more I'm convinced that there are 7.4 billion different types of people on the planet.

"Empathy" is a word that is often used in design. What does it mean to design with empathy for billions of people? As technologists grow more aware of the exclusion that their products might create, there's a renewed focus on increasing empathy. This word, like the word "inclusion," has many different interpretations.

Some design methods for empathy focus on simply talking to people. For example, introducing yourself to people on the street and asking them for feedback on an idea. Other approaches emphasize ways to question your own assumptions and abstractly consider how other people might feel about a problem. Some teams have extensive research departments where sociologists, anthropologists, and psychologists conduct studies with people. Most of these approaches don't scale to the level of billions of people.

Various cultures describe the concept of empathy in different ways. In the context of designing technology, there are two Mandarin Chinese descriptions of empathy that are particularly useful, shown here next to an English translation of each individual character.

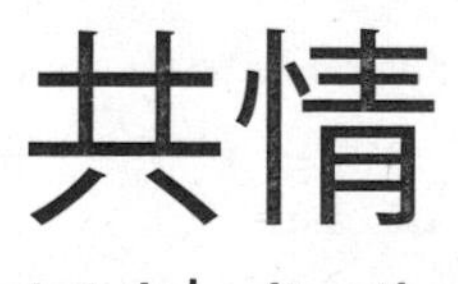

Figure 7.9
Two Mandarin descriptions of empathy.

Empathy, "to reason with the heart." Empathy, to sense "the total situation." Building solutions on a global scale requires designers to do both, equally well. It's making sense through the greater context and the relationships that lie therein, incorporating both the breadth and depth of humanity. This is how empathy will evolve to meet the challenges of designing for 7.4 billion people.

Inclusion requires us to shift our assumptions about who receives the things we design. It starts with curiosity and observation. Analyzing massive amounts of data can amplify new categories of problems to solve. It can help us iterate through ideas and optimize our solutions more quickly. And when we craft those solutions we have a responsibility to make sure they work well for people, functionally and emotionally. These are nuances that come from real-life, human-to-human conversations.

Not everyone is an anthropologist. We aren't all mathematicians. Not everyone is great at putting aside their cultural preconceptions. But we can all learn how to seek out excluded perspectives and let them supersede our own.

▪ ▪ ▪

TAKEAWAYS: TESTING OUR ASSUMPTIONS ABOUT HUMAN BEINGS

Exclusion habits

- Designing for a mythical average human that presumes 80% majority and 20% minority use cases.
- Assuming that people will adapt themselves to make a solution work.

How to shift toward inclusion

- Create a baseline assessment of how well your solution works for excluded communities, especially people with disabilities.
- Create a persona spectrum that is specific to the problem you're trying to solve.
- Build thick data with the people who face the greatest mismatches in using your solutions. Learn from their expertise with thoughtful observation and inquiry.
- Balance big data with thick data. Use big data as a heat map to reveal key mismatches between people and your solution. Use thick data to investigate the reasons behind these mismatches and gain insight into better solutions.
- Build one-size-fits-one solutions to fit people who likely face the greatest mismatches when using your solutions. Extend the solution to people who face similar mismatches on a temporary and situational basis. Focus on the shared reasons why each of these groups wants to participate in your solution.

8 LOVE STORIES

How inclusion drives innovative outcomes.

How many objects in your home are the result of inclusive design?

Maybe it's the adjustable chair at your desk. The keyboard to your computer. Your smartphone touchscreen. Your reading glasses. These, and many more objects, are the descendants of a long history of innovations that were made to remedy exclusion.

Many assistive solutions that were originally marketed to people with disabilities eventually found mainstream potential. As technology improves, functionality gets better and market opportunities expand. In turn, businesses that recognize these opportunities are more likely to invest in making the design of that product highly usable and beautiful.

This remains true for a number of rising technologies. Speech recognition and voice commanding were, in part, initially designed for people with disabilities as a way to interact with computers without using a keyboard, mouse, or screen. Today, millions of people are

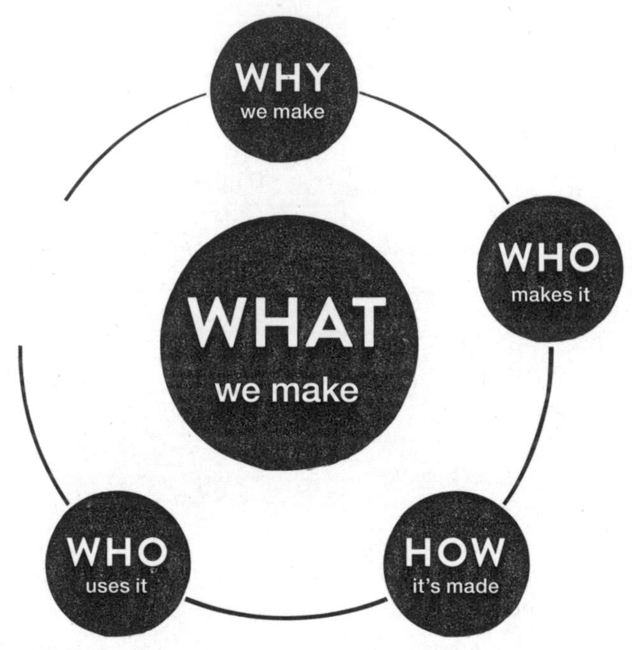

Figure 8.1
Problem solvers are often eager to start at this point on the cycle, jumping to solutions. This makes it a powerful way to make the business case for shifting toward inclusion.

talking to their cars and household devices to get directions, check the weather, and order pizzas using the same core technologies.

The deep exclusion habit in *what we make* is the fixed meaning we assign to objects. A classic creative thinking exercise is to imagine multiple ways of using a typical brick. A brick could be used to build a wall. This fits the expected context and application of a brick.

A brick can also be utilized to hold a door open in the summer. It can be heated and used to cook a meal. Or ground into dust and used as sand. When we change the shape, context, or purpose of an object, it can take on a new meaning.

By stretching our assumptions about the purpose of an object or environment, we can explore how that solution flexes to be whatever a person needs it to be. The signature trait of an inclusive solution is how it adapts to fit each unique person.

This stretch of creativity can be particularly challenging for problem solvers who think about excluded communities and accessibility in fixed ways. A brick is a tool for building walls. A screen reader is a tool for someone who's blind. Video captioning is a tool for people who are deaf. Accessibility products are often thought of as fixed and specialized solutions for people with disabilities.

As a result, accessible solutions often lack thoughtful design. Physically assistive objects like canes, wheelchairs, prosthetic limbs, or support braces can resemble the cold, medical objects of a surgeon's operating room: steel metal bars, rough materials, industrial-grade plastics, and lifeless colors. Everything about the design can feel like a cue to the user, and to others around them, that they are a patient in need of help.

The same is often true in the creation of digital products. When a solution is treated as "for disability" or "for accessibility," there's often little or no attention paid to the design. A solution might meet all of its functional requirements but still lead to emotional or aesthetic mismatches that can be equally alienating.

Assistive devices, in particular, often play an intimate role in a person's life. They become much more than just objects. They can be an extension of a person's independence, beauty, strength, and connection with the world.

The fixed way of thinking about assistive objects often triggers a few common questions before leaders are willing to invest in inclusion:

- What is the business case for it?
- What is the return on investment?
- How can you prove that it works?

Before we dive into business justifications, it's helpful to study tangible examples of innovations that grew out of mismatched designs. A person who has experienced a great mismatch is likely to take an object that was intended for one purpose and use it in new ways. A desire to participate demands that they work creatively with what's available to resolve the mismatches for themselves.

Here are some of my favorite examples of inclusive design and the stories behind them.

THE FLEXIBLE STRAW

Joseph Friedman was at an ice cream shop with his young daughter. Sitting at the counter, she was having a hard time drinking her milkshake through the straight paper straw without spilling the drink. Friedman inserted a screw into the straw and tightly wrapped a wire around it to make a flexible joint.

That bend in the straw made it possible for his daughter to comfortably enjoy her beverage. But it also works well for anyone who's unable to hold a glass to their mouth, or reclined in bed from illness or injury. It also benefits anyone who's reclined on a beach, enjoying their favorite vacation beverage.

TYPEWRITERS AND KEYBOARDS

One of the first typing machines was invented with a woman named Countess Carolina Fantoni da Fivizzano, in the early 1800s. She was the friend and lover, according to some sources, of Italian inventor Pellegrino Turri. The countess slowly lost her vision at a time when the only way for someone who was blind to send letters was by dictating their note to another person, who would transcribe the message to paper.

In order to keep their communications private, the Countess and Turri invented a machine that could be used to write notes by pressing a key for a single letter, raising a metal arm to press each letter into carbon paper. This invention made writing accessible to people who are blind, but many derivatives have evolved over the past two centuries into the modern-day keyboards for computing and mobile devices.

FINGERWORKS

Wayne Westerman wanted to create a method of interacting with a computer that required no force in the hand. He was motivated, in part, by his own severe case of carpal tunnel syndrome. His company, FingerWorks, developed a way to replace a keyboard with a touchpad for each hand.

They initially marketed their invention to people with hand disabilities and repetitive-strain injuries to their arms. The company had a base of passionate customers who came to depend on FingerWorks products in their daily computer usage. There was also an increase in customers who were interested in the design as an easier way to navigate their computer, regardless of their abilities or limitations.

FingerWorks sold their technology to Apple in 2005, enabling the tech giant to build their first gesture-controlled multitouch interface, the iPhone. Westerman is listed on a 2007 iPhone patent.[1] However, in the process, the original FingerWorks product was discontinued.

PILLPACK

The childproof cap on prescription bottles was invented in the 1960s, about the same time that manufacturers started to apply candy coating to pills of aspirin. They needed a way to ensure that children couldn't access the bottles.

However, childproof caps are hard for everyone to use, not just children. People with limited dexterity of their hands, many of them due to age or injury, have a particularly difficult time opening a childproof bottle. Further, for people who take multiple medications, the risk of mistaking medication can be serious, even deadly, and a curved surface of a pill bottle increases this risk by making it difficult to read instructions on the label.

PillPack founder and second-generation pharmacist T. J. Parker and his team partnered with IDEO to design a better way to deliver prescriptions. They focused on people who faced the greatest mismatches in using medication. Some of them were long-term cancer patients who took over a dozen different medications in one week.

Their final solution wasn't simply a more accessible pill bottle. They reimagined the prescription delivery service. A patient can order their prescriptions through PillPack and their medication arrives in presorted packages. So rather than the patient needing to organize their medications manually, they simply remove one small pouch that contains the right medication for the right time of day.

For the 30 million people who needed to take more than five prescription medications in one day, this solution drastically improves safety and convenience. PillPack is now working to improve the system of interactions between patients, pharmacies, doctors, and insurance companies, with the launch of their comprehensive service Pharmacy OS.

SONIFICATION OF THE STARS

Wanda Díaz-Merced is an astronomer, born and raised in Gurabo, Puerto Rico. A passionate math and science student, she and her sister dreamed of riding in a space shuttle. While studying astronomy, she started to lose her eyesight. For an astronomer, one of the key ways to study data from astronomical events is through visual illustrations. As her eyesight changed, Díaz-Merced worried that she wouldn't be able to continue to practice and grow in her profession.[2]

She started using a technique called sonification, which makes it possible to "listen" to the stars by turning the data from stellar radiation into audio files. She matches the patterns of the data to pitches of sound. During her PhD research she explored new and better techniques for sonification of astronomical data. She wrote:

> Sound offers a way to increase sensitivity to visually ambiguous events embedded in the kinds of data space scientists and astronomers analyze. Radio waves are conveyed by drums, x-rays by the harpsichord, and so on.[3]

Díaz-Merced continues to practice astronomy at a world-class level. Sonification also improves access to astronomical data for a much broader group of people. In particular, for any astronomer who is unable to distinguish the colors used in a data visualization or anyone who's losing their vision due to age or injury. And for practitioners in any data-oriented field, using the combination of audio and visual information gives them greater access to the nuances of their data.

EMAIL

Vint Cerf is best known as a father of the Internet. He's considered one of the primary architects of its early development. In 2002, Cerf authored "The Internet Is for Everyone," a summary of what makes Internet access a human right, key threats to that access, and a plea for the Internet to be engineered in ways that make it usable by anyone in the world. To this day, Cerf asserts the importance of accessibility: "It's almost criminal that programmers have not had their feet held to the fire to build interfaces that are accommodating for people [with disabilities]."[4]

He also created some of the earliest protocols for email. Cerf is hard of hearing and his wife is deaf. At the time, they couldn't use a telephone to communicate. His work on email started, in part, as a way for him and his wife to stay connected when they weren't together in the same room.

Email, of course, is today nearly as ubiquitous as the Internet. Along with closed captioning and subtitles, it has become a key technology for people who are deaf, have hearing loss, or are simply separated by time and space.

MORGAN'S INSPIRATION ISLAND

In the blazing summer heat of San Antonio, waterparks are an important place for children and their families to cool down and play. One waterpark is named after Morgan Hartman, who, with her parents Gordon and Maggie, built this "park of inclusion." They were inspired to do so after years of being unable to find great play spaces for Morgan, who has cognitive and physical disabilities. They set out to create a place where children with and without disabilities could play together in a barrier-free environment.

There are a few things that make Morgan's Inspiration Island extraordinary, beyond its colorful water features and accessible design.

First, the Hartmans included a wide range of people with disabilities and specialists in creating the park. Second, they developed ways to adjust water temperature in real time so it can be personalized to a child's sensory needs. Third, they worked with engineers at the University of Pittsburgh and the U.S. Department of Veterans Affairs to stock the park with motorized wheelchairs that are powered by compressed air, a design called the PneuChair.[5] The chairs are safe for water play and only take 10 minutes to recharge.[6] And lastly, the park is free of charge to anyone with a disability.

Many of these examples are love stories. In fact, love is a common trait in the creation of inclusive solutions. Some stories center around a person who was personally affected by a mismatched design. Their love for a profession or lifelong passion for an activity put a sharp focus on how it could be improved. Then they worked directly on building a better solution.

Other stories come from a mismatch that loved ones faced when something interrupted their connection to each other.

In all cases, people worked with their intimate understanding of exclusion, and with the participation of excluded communities, to design a solution that went on to benefit a wider group of people. There are many more stories of inclusive design, some that reach massive commercial success and others that are working quietly in the background of everyday life. Drawing from these examples, here are four ways to start building a business justification for inclusion:

1. Customer Engagement and Contribution

Engagement with a product increases when it's easier to use. The key to this business justification is to demonstrate exactly how mismatched designs are affecting real customers. Work with excluded communities to record the challenges they face when using your product. Detail the kinds of hacks that they use to make the product work. Make it crystal-clear and tangible. Present these stories to leadership teams as obstacles to customer engagement and explain how removing those specific mismatches can reduce the friction for many more customers.

Another type of engagement happens when customers contribute to the development of your product. Most types of technology and how they're made are largely mysteries to the general public. Yet these solutions play a primary role in people's daily lives. Their inclusion can mean more to them than you realize. Meeting in person and incorporating their input can increase their sense of belonging with your brand and product. Listen carefully to people and pay attention to the emotional connections. Share the stories of how they influenced your design decisions.

2. Growing a Larger Customer Base

It can seem counterintuitive to start with a sharp focus on excluded communities. The strength of this approach is that it outlines clear constraints, helping teams build a deep understanding of how to connect with a wider target audience.

In a similar way, aspirational brands commonly focus on an elite community to build mass-market products. For example, working with world-class athletes to build a new footwear line. Or partnering with blockbuster filmmakers to explore augmented reality experiences. Stronger constraints can push designers and engineers to innovate. The key is to do so in a way that translates to a broader market by finding ways to make the solution relevant to a general audience.

Another way to build a business case for inclusion based on market size is through the persona spectrum. Quantify the number of people whose abilities lie within the permanent, temporary, and situational categories of a spectrum of exclusion. If more is better, this argument is one way to present inclusive design as a significant market opportunity.

3. Innovation and Differentiation

Leaders are often surprised by how inclusion can fuel innovation. Accessible solutions, in particular, have a history of seeding innovations that benefit a broader audience. This is for two reasons.

First, many companies have decades' worth of ideas and prototypes that they've developed but never released into the world. These solutions often sit unused, like a pantry of ingredients in a kitchen. Then, with a shift in perspective or context, an ingredient that was originally intended for one purpose becomes useful in a new way. When teams shift toward inclusion they discover new solutions, but also new ways to make use of existing but unused solutions.

For example, screen contrast. Computer displays can be adjusted to increase the contrast in colors between different elements, like text and the background it sits against. This feature was vitally important to people with types of low vision that make it difficult to distinguish between objects on a screen if their colors are similar.

Screen contrast became relevant to a much wider group of people once mobile phones arrived. Suddenly, anyone who was using a phone in bright sunlight had a hard time reading the information on the screen. As smartphones evolved, they used existing screen contrast technologies to automatically adjust a screen and make it readable outdoors. A new problem created new relevance for an existing technology, improving the product for a wider group of customers.

Second, innovation is amplified by new perspectives. Through inclusive design methods, a team learns how to open their process to include people with complementary ability biases and new kinds of expertise. Sometimes the person who can have the greatest impact on a solution is someone who has relevant expertise but isn't deeply involved in the technical intricacies.

Many inclusive innovations don't require a dramatic reinvention of technology. They don't require tearing down existing solutions to create new ones. Often, it's just applying a new lens to the resources that already exist, and forming new combinations of existing solutions. It starts by employing new perspectives to reframe the problems we aim to solve.

4. Avoiding the High Cost of Retrofitting Inclusion

Many teams and companies treat accessibility and inclusion as an add-on, something to consider only in the final stages of completing a product. It's the unfortunate result of the bell-curved belief in an average human. A team that treats inclusion as an afterthought focuses first on people they imagine are most like themselves, especially in ability and in cognitive and societal preferences. They commonly believe that this group of people represents the majority of their audience.

Conversely, they will assume that demographically underrepresented groups of people, like people with disabilities, are edge cases, a small percentage of the population that doesn't represent large opportunities for revenue. This is simply a myth. The myth of the minority user.

After decades of building products with an average-human mindset, there is a lot of neglected accessibility work that needs to be addressed. While it's hard to gauge exactly what percentage of all websites are inaccessible, a growing number of accessibility audit firms will tell you that their clients are often surprised at how many of their websites and digital products don't meet legal standards for accessibility. It can take huge resource investments to fix these basic issues. This is the high cost of treating accessibility as an afterthought.

Another way to quantify the cost of retrofitting inclusion is the number of lawsuits and public relations missteps companies face when they release discriminatory solutions. From aggressive startups to careful tech incumbents, there are plenty of examples of how costly it can be to neglect issues of inclusion during the product development process.

Cathy O'Neil, author of *Weapons of Math Destruction*, does an excellent job of detailing the potential dark depths of human biases that are baked into technology.[7] Beyond the accidental chat bot that's trained to spew hate speech, or a selfie filter that resembles insensitive racial caricatures, O'Neil provides examples of how machine learning is amplifying the cycle of exclusion on a massive scale, from predictive policing to the algorithms that determine what advertisements show up in your phone. The cycle of exclusion is amplified when machines are coded through human bias.

Ideally, every new product or project would consider inclusive design from the beginning, as a way to proactively save time and resources. There would be no retrofitting required. More feasibly, inclusive design needs to happen when and where it can, while always pushing to happen earlier in the development process.

The best way to do this is to weave inclusive design methods throughout the entire process of developing a solution. There is no one-size-fits-all approach to practicing inclusive design. Each company needs to employ a combination of methods that complements their existing processes.

BEAUTIFUL OBJECTS, HUMAN EMOTIONS

The cycle of exclusion culminates with the solutions that we introduce into the world. *What we make* is a powerful starting point in the cycle of exclusion. It can spark the imagination. It's the point in the cycle that embodies our excitement for innovation.

Introducing a new technology into society can have a profound impact on how people feel and behave. Sometimes it's an intense obsession, like Pokémon Go. At other times, it's a lukewarm affection that grows into a deep dependency, such as touchscreen smartphones. The relationships between humans and objects are filled with emotions.

Being rejected by exclusionary designs can lead to negative emotions. In contrast, a design that fits a person's unique body and mind can have a positive emotional impact. Whether it's the beauty of listening to the sounds of a distant star or the sense of adventure that comes from making new friends in a water park, the mark of a successful inclusive solution is that it's both a functional and an emotional fit.

In many ways, people are slower to change than technology. Building inclusive solutions doesn't mean that everyone has to personally embrace the values of inclusion. As we discussed at the opening of this book, inclusion can easily be mistaken for being a nice person. What if, instead, the skills of creating inclusive solutions were the measure of a successful engineer, teacher, civic leader, or designer?

What if, rather than trying to change how problem solvers feel about inclusion, we could build inclusivity by changing how we create solutions? Could that be a faster path to a more inclusive society?

What we choose to make shapes the future of who can participate in and contribute to society. Mismatches can help us stretch our thinking far beyond technology for the sake of technology. A brick is more than a brick. The objects that assist us are more than accessories. Inclusive methods can help ensure that in our pursuit of innovative solutions we are also making solutions that are humane.

■ ■ ■

TAKEAWAYS: HOW INCLUSION DRIVES INNOVATIVE OUTCOMES

Exclusion habits

- A limited willingness to imagine how a solution in one context can adapt to provide new kinds of value in a different context.
- Only focusing on functional outcomes and ignoring emotional mismatches.

How to shift toward inclusion

- Assess the greatest mismatches in what you make. What barriers currently exist, and how would adapting those barriers open access to more people?
- Explore how a solution can adapt to be whatever a person needs it to be. Focus on creating flexible systems that fit people in unique ways as they move from one environment to the next.
- Seek out inclusive-design love stories that already exist in your business. How are people using and adapting your solutions as a way to connect with the people and activities that they love?
- Build a business case for inclusion based on how it supports:
  - Customer engagement and contribution,
  - Growing a larger customer base,
  - Innovation and differentiation,
  - Avoiding the high cost of retrofitting inclusion.

9 INCLUSION IS DESIGNING THE FUTURE

Why inclusion matters.

We've taken a closer look at each element of the cycle and how we might shift exclusion habits toward a more inclusive path. Now, can we again ask ourselves, why do we make?

Why would we sign up for the hard work of building inclusion without the guarantee of success? Why would we fight the inertia of a cycle of exclusion that's been spinning for generations?

Certainly, there are business justifications like gaining new market share, creating better customer experiences, and operating in more efficient ways. There are opportunities to connect teams to a meaningful purpose for their work and a new way of thinking about the people who interact with their solutions.

There are professional reasons why inclusion matters. It expands our own thinking about problems that are worth solving. It sparks our creativity to think in new ways, in partnership with new people.

And there's our collective future. A future that is built on the choices that we make today, to create great solutions that connect people to each other and to opportunities in the places where they live. Some designers will make choices that reach millions of people and will endure for many years. If nothing else, I hope this book illustrates the weight of that privilege and opportunity.

But there is one reason for inclusion that transcends all other justifications: uncertainty.

So many of the strategies that have led us to exclusion were about avoiding uncertainty. As children, we protect our games from disruption and uninvited intrusions. As designers and engineers, we

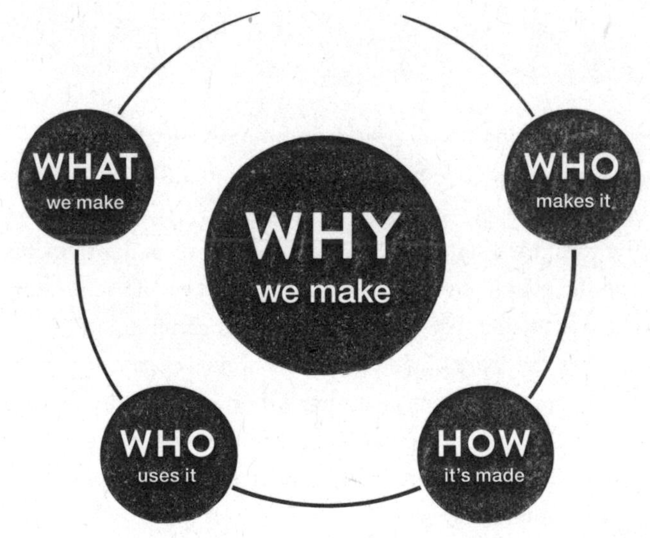

Figure 9.1
We can shift toward inclusion when we apply inclusive design principles to any element of the cycle of exclusion. This includes reframing why it matters to our own lives, our solutions, and our society.

use mathematical models to homogenize the people we design for. As architects, we wipe away existing patterns of familiarity with a mandate to create order out of perceived chaos.

Yet we now face greater uncertainty than ever before. Not just in the ways digital technologies permeate our societies, but in our connections with each other.

OUR FUTURE SELVES

My preteen daughter had this to say when I asked her to describe design.

> It's a series of movements, powered by the human body, hand, or foot, or any part of the body. It's also a thought in your mind that can be written, said, or drawn out. Everything that you make, from a knot in your hair to a satellite, is designed. There's a lot of really good ideas out there.

She tells me that in eight years there will be true hoverboards and medicines that prevent people from dying at a young age. If there was one problem that she could solve today, it would be making sure that everyone has a home to keep them safe and healthy.

I can't help but consider her answers in the context of my father's life. When he was a child, his family was detained at Angel Island as part of the Chinese Exclusion Act, passed by the U.S. Congress of 1882. The Act initially prohibited all Chinese laborers from entering the U.S. for 10 years, and was extended in 1924 to all Chinese immigrants. It wasn't fully repealed until 1943.

Shortly after they were released from Angel Island, his father and toddler-aged sister died of respiratory illnesses. My grandmother had to make a life for herself and my father in a new country. It's an incredibly difficult time for him to recount. Too many sad memories, he says.

As kids, my sister and I would follow him through the streets of San Francisco. We'd explore the hills of Chinatown, where our grandmother ran her business trading Tahitian imports and exports. In particular, we loved the YMCA yard where he played with friends and the basement after-school center where he stayed extra hours to study Cantonese. They lived in the back of their shop for years, until my grandmother saved enough money to buy a home in Oakland, across San Francisco Bay.

On our final trip together to the city we walked the path along Stockton Street looking for the old shop and found, in its place, a fifteen-foot wall that spanned the entire block. Signs across the wall announced the upcoming construction of a new transit station on the site. After that, he never wanted to go back to the city again.

Exclusion and inclusion shaped our family in deep ways. Gaining access to American society was one level of inclusion. Gradually being able to participate in that society was another level. Perhaps, also, being able to bring his own kids back to that first home was another kind of inclusion. One that represented a sense of belonging.

The same things that motivate my daughter also motivated my family over ninety years ago: health, safety, a home, and the desire to create. Making something that contributes to society in positive ways. I hope she's right. I hope the near future holds the promise that she imagines. And that she sees herself as an important contributor to that future.

With my father, I recognize how my future body will change. My hearing will fade by the time I'm seventy. My eyesight will cloud over and grow noticeably worse at night. My memory will remain sharp on some details, and I'll start to forget others. My arms and legs will lose their strength. How will the design of my environment ensure that I can participate in society, rather than becoming socially isolated with age?

This might be the most immediate kind of uncertainty that we each face. Our ever-changing abilities. Many of us are temporarily able-bodied and will face new kinds of exclusion as we age. When we design for inclusion we are designing for our future selves. Not just for the changes in our bodies but also our ability to contribute to society. It's designing how the next generation will treat and care for us. It's making solutions to uphold the human connections that are most important in our lives. Our dignity, health, safety, and sense of being at home.

FUTURE OF TECHNOLOGY

Another area of uncertainty is how the next generation of technology will unfold.

The evolution of technological ages can be framed in multiple ways. I like to think about them in terms of the relationship between humans and machines. The industrial age was initially fueled by the mechanization of production that replaced some types of human labor. People consumed the products of industrialization, from mass-manufactured clothing to food to cars. Each object played one or two specific roles in a person's life.

The information age gave rise to human-computer interactions, where people met machines through a digital interface. Although these interfaces have grown complex and play many roles in our lives, they are still largely the product of a fixed set of choices determined by an engineer or designer. They are hard-coded to perform certain tasks and follow certain rules. Their behavior is planned, to a large degree. And the people who interact with these interfaces have to adapt themselves to consume information in the predetermined ways that a program can deliver it.

We are now feeling the pressure of transition to a new age of technology. Most companies are scrambling to make use of the overwhelming amounts of data being collected from people using digital interfaces, nearly 1.7 megabytes of new data per human per second.[1] Today, about 90% of that data is unstructured, meaning it doesn't follow a predefined data model and takes a lot of human manipulation to turn it into useful information. The small group of people with the elite skills to craft this data, and the machine learning models that consume it, are more like specialized artisans than traditional software engineers.

How people interact with these new systems is also changing.

In the face of this change there's a lot of experimentation happening. There's great enthusiasm for the ways that machine learning models can generate, refine, and optimize better designs. Data from these models can persuade individual leaders to make decisions using more than their personal biases and preferences. Many pioneers in data-driven product development will tell you that the customer, through data, is always right and that it's best to move fast and adapt through real-time customer feedback.

Yet how do we avoid overdependence on mathematical models to reduce uncertainty, an overdependence that will lead us to cycles of exclusion similar to Quetelet's normal curve? Can big data and machine learning algorithms respect human individuality, or will they just move all of us toward the center of the curve? Our enthusiasm for data analysis could easily override our willingness to understand its limits. Most designers and engineers are just starting to explore what these models are great at, and when their results are deceptively wrong.

Questions like these are important because the rise in machine learning will form the backbone of new human-computer interactions. When you're talking with an artificial intelligence to order a movie or send a message to a friend overseas. When you apply for a student loan, or a job. Or when your thermostat adjusts the temperature in your home. These machines are programmed to decide how to respond to you without human consultation. They are making recommendations to product teams on how to improve products using data amassed from millions of people.

These technologies aren't new. Many of them have been around for decades. The explosion of data and specialized hardware are fueling a new surge in machine learning and artificial intelligence solutions. So now, more than ever, the balance of thick data and big data is important. In particular, thick data that helps us understand the complexity of human behavior. The underlying cultural and personal reasons why people do what they do. And a great place to start understanding why they do is through the inclusion of excluded communities.

For all the debate around the benefits and dangers of the various forms of learning machines, one trait makes this technology particularly interesting to inclusive design: adaptation.

A machine that can flex itself to the unique needs of one person could be deeply personal to millions of people in infinite ways. It's easy to understand why there's an excited frenzy of innovation in these new classes of technology.

However, there are emotional consequences. Think back to those kindergarten classrooms and the sting of rejection. Being personal and adaptive cuts multiple ways. When something fits you beautifully, and feels like a true match to your individual needs, it can be a deeply positive experience. Conversely, when a solution excludes you, it can feel deeply personal as well. When a machine makes the decision on who to include or exclude, how will that feel?

For every new area where technology is introduced as a facilitator of human interactions, what extra degree of responsibility must be built into that machine? How do we hold that machine and its creators accountable for exclusionary or harmful outcomes? Will we even recognize how those consequences are shaping the world around us if the general public doesn't have a basic understanding of how these solutions work? Will already marginalized communities become even more invisible?

These are not assertions. They are genuine questions. These are the choices that every problem solver, engineer, architect, designer, lawmaker, entrepreneur, leader, and more should be asking as digital technologies permeate every area of society.

These moments of technological transition are the ideal time to introduce inclusive design. We can engineer these new models to ensure that they don't lead to exclusionary design for the nonexistent average human. Without inclusion at the heart of the AI age, we risk amplifying the cycle of exclusion on a massive scale. It won't just be perpetuated by human beings. It will be accelerated by self-directed machines that are simply reproducing the intentions, biases, and preferences of their human creators.

INCLUSIVE GROWTH AND HUMAN POTENTIAL

Perhaps the greatest uncertainty is how global economies, governments, and businesses will respond to the changes being driven by the next generation of technologies, and how their reactions will shape design decisions. The approach to inclusion in this book is heavily based in U.S.-centric definitions of inclusion, accessibility, and disability. Globally, there are varied approaches to inclusion that are rooted in cultural factors, such as language or the categories that are used to differentiate between groups of people. For many countries, shifting cycles of exclusion toward inclusion is, ultimately, an economic shift. A profound one.

Why does growth matter? Who will lead it? Who will it benefit, and how will we get there? On a global scale, there's a rising conversation about inclusive growth. Inclusive growth is defined as "economic growth that creates opportunity for all segments of the population and distributes the dividends of increased prosperity, both in monetary and non-monetary terms, fairly across society."[2]

The World Economic Forum (WEF) further describes inclusive growth as a positive or negative feedback loop between economic growth and social inequality.

> There is mounting evidence that inequality has a statistically significant negative impact on growth, and that reducing inequality can enhance and strengthen the resilience of growth. For example, if the income share of the top 20% increases, GDP growth tends to decline.[3]

The WEF predicts that at least 5 million jobs in the world's leading economies will be lost to tech by 2020.[4] It's easy to imagine how such large-scale changes in the global workforce could lead to greater social inequality. Instead, how might these changes become opportunities to reduce inequalities and support inclusive growth? Inclusive design could be one way to start. Not just for the products and environments of our world, but also for the systems that connect them.

The landscape of employable skills and talents is also shifting. In their "Future of Jobs" report, the WEF lists its prediction for the top ten skills for future jobs:[5]

1. Complex problem solving,
2. Critical thinking,
3. Creativity,
4. People management,
5. Coordinating with others,
6. Emotional intelligence,
7. Judgment and decision making,
8. Service orientation,
9. Negotiation,
10. Cognitive flexibility.

These skills are focused on making sense of uncertainty; on taking multiple approaches to framing a problem worth solving and finding solutions that adapt to fit unique situations. These are the talents of building connections with human beings.

We have a lot of work to do to meet these talent demands, including new ways to recognize and remedy exclusion. If we simply carry forward today's solutions for creating designs, writing code, or communicating with people, we risk making these roles even more exclusive than they are today.

Inclusive design is simply good design for the digital age. Shifting *who* can contribute their talents to creating solutions will, in turn, shift what we make, how it's made, and who benefits from those designs. We are all the beneficiaries of inclusive design over the course of our lifetimes.

Design is essential to inclusive growth. It closes the gaps in the interactions between people and the world around them. We need solutions for human environments that adapt to fit individual needs, diverse bodies, and diverse minds.

Figure 9.2
The three principles of inclusive design.

Returning to the central question of this book: If design is the source of exclusion, can it also be the remedy? Yes. When the principles of inclusive design are applied to any element of the exclusionary cycle, it creates inclusion.

We can interrupt our habits, even if just for a moment, to recognize the exclusion around us. And invite people who have complementary ability biases to be co-designers in our solutions.

We can learn from people whose exclusion expertise could be the key to unlocking some of our toughest challenges. Our one-size-fits-one solutions can extend to serve many more people. When we bring it all together, we create a diversity of ways to participate so that everyone has a sense of belonging.

As a problem solver you can shift the cycle toward inclusion, one choice at a time. With each design, you shape who can contribute their talents to society. Their contributions, in turn, will shape the future for all of us.

▪ ▪ ▪

TAKEAWAYS: WHY INCLUSION MATTERS

Exclusion habits

- Lacking clarity and agreement on what inclusion is and why it matters.

How to shift toward inclusion

- Use the cycle of exclusion to assess where you are today and where to start.
- Apply the principles of inclusive design to any element of the cycle of exclusion.
- Integrate inclusive design methods within your team to build a purposeful culture where people can do their best work.

NOTES

Chapter 1

1. Martin Verni, “Designer Spotlight—Susan Goltsman and the Emergence of Inclusive Design,” January 20, 2016, https://goric.com/susan-goltsman-inclusive-design/.

Chapter 2

1. Vivian Gussin Paley, *You Can’t Say You Can’t Play* (Cambridge, MA: Harvard University Press, 1993), 20–22.

2. Inclusive: A Microsoft Design Toolkit, Subject Matter Expert Video Series, 2016, www.mismatch.design.

3. Inclusive: A Microsoft Design Toolkit, Subject Matter Expert Video Series, 2016, www.mismatch.design.

4. Marshall McLuhan, *Understanding Media: The Extensions of Man* (1964; Cambridge, MA: MIT Press, 1994), xxi.

5. Mandal Ananya, “Color Blindness Prevalence,” *Health*, September 2013.

Chapter 3

1. The World Bank, *World Report on Disability: Main Report (English)* (Washington, DC: World Bank, 2011).

2. Vivian Gussin Paley, *You Can't Say You Can't Play* (Cambridge, MA: Harvard University Press, 1993), 22.

3. Cornell University's Online Resource for Disability Statistics, http://www.disabilitystatistics.org/.

4. US Bureau of Labor Statistics, "Labor Force Statistics from the Current Population Survey; Databases, Tables & Calculators by Subject," data extracted on January 22, 2018.

5. Resources for learning about these policies are listed as Suggested Reading at the end of this book.

6. To learn more, visit the work of Kipling Williams at Purdue, Ethan Kross at University of Michigan, Naomi Eisenberger and Matt Lieberman at UCLA, Amanda Harrist at Oklahoma State University, Nathan DeWall at the University of Kentucky, and Ronald Rohner who founded the Center for the Study of Interpersonal Acceptance and Rejection at the University of Connecticut.

7. N. Eisenberger, M. Lieberman, and K. Williams, "Does Rejection Hurt? An fMRI Study of Social Exclusion," *Science* 302, no. 5643 (October 2003), 290–292.

Chapter 4

1. World Bank, *World Report on Disability: Main Report (English)* (Washington, DC: World Bank, 2011).

2. Quoted in *Inclusive*, a short film by Microsoft Design; www.mismatchmedia.com.

3. Quoted in *Inclusive*, a short film by Microsoft Design; www.mismatchmedia.com.

4. Inclusive: A Microsoft Design Toolkit, Microsoft Design, 2015.

5. More resources on disability and accessibility policies are listed in Suggested Reading.

6. Ireland's Disability Act of 2005; Centre for Excellence in Universal Design.

7. The seven principles of Universal Design, published in 1997 by Ronald Mace and a team of architects, designers, and engineers when he was at North Carolina State University:

1. Equitable use
2. Flexibility in use
3. Simple and intuitive use
4. Perceptible information
5. Tolerance and error
6. Low physical effort
7. Size and space for approach and use

8. More resources on inclusive design, accessibility, and universal design are available at http://www.mismatch.design.

9. More resources on accessibility are listed in Suggested Reading at the back of this book.

Chapter 5

1. National Council of Architectural Registration Boards, "Timeline to Licensure," in "NCARB by the Numbers," 2016.

2. Amy Arnold and Brian Conway, *Michigan Modern: Design that Shaped America* (Layton, UT: Gibbs Smith, 2016).

3. "Detroit (city), Michigan," State & County QuickFacts, United States Census Bureau. Retrieved January 2017.

4. Campbell Gibson and Kay Jung, "Historical Census Statistics on Population Totals by Race, 1790 to 1990, and by Hispanic Origin, 1970 to 1990, for Large

Cities and Other Urban Places in the United States," table 23, "Michigan—Race and Hispanic Origin for Selected Large Cities and Other Places: Earliest Census to 1990," United States Census Bureau, February 2005.

5. U.S. Census Bureau, "American Community Survey 1-Year Estimates," 2016. Retrieved from Census Reporter Profile page for Detroit, MI.

6. Toni L. Griffin and Esther Yang, "Inclusion in Architecture September 2015," report from the Anne Spitzer School of Architecture, City College of New York.

Chapter 6

1. Katherine Shaver, "Female Dummy Makes Her Mark on Male-Dominated Crash Tests," *Washington Post*, March 25, 2012.

2. D. Bose, M. Segui-Gomez, and J. R. Crandall, "Vulnerability of Female Drivers Involved in Motor Vehicle Crashes: An Analysis of US Population at Risk," *American Journal of Public Health* 101, no. 12 (2011), 2368–2373.

3. Margaret Burnett, "GenderMag: A Method for Evaluating Software's Gender Inclusiveness," *Interacting with Computers, The Interdisciplinary Journal of Human-Computer Interaction* 28, no. 6 (November 2016).

4. Margaret Burnett, Robin Counts, Ronette Lawrence, and Hannah Hanson, "Gender HCI and Microsoft: Highlights from a Longitudinal Study," IEEE Symposium on Visual Languages and Human-Centric Computing, October 2017, pp. 139–143.

Chapter 7

1. Todd Rose, *The End of Average: How We Succeed in a World That Values Sameness* (New York: HarperOne, 2015).

2. Lambert Adolphe Jacques Quetelet, *A Treatise on Man and the Development of His Faculties* (1835; Cambridge: Cambridge University Press, 2013), 99.

3. J. M. Juran, *Architect of Quality* (New York: McGraw-Hill, 2004).

4. Rose, *The End of Average*.

5. Alden Whitman, "Margaret Mead Is Dead of Cancer at 76," *New York Times*, November 16, 1978.

6. World Health Organization; "Fact Sheet on Deafness and Hearing Loss," February 2017.

7. "American Deaf Culture," Laurent Clerc National Deaf Education Center, Gallaudet University, http://www3.gallaudet.edu/clerc-center/info-to-go/deaf-culture/american-deaf-culture.html.

8. Dominic Barton, Jonathan Woetzel, Jeongmin Seong, and Qinzheng Tian, "Artificial Intelligence: Implications for China," McKinsey Global Institute, April 2017.

9. Jessica Qiao, Juliana Yu, and Frank Wang, "IDC Announces Top Predictions for China's Internet Industry in 2017," press release, March 2017, https://www.idc.com/getdoc.jsp?containerId=prCHE42353017.

Chapter 8

1. Brian Merchant, *The One Device: The Secret History of the iPhone* (New York: Little, Brown, 2017), 81.

2. "How Can We Hear the Stars?," Guy Raz interviews Wanda Díaz-Merced, NPR *TED Radio Hour*, January 2017.

3. Wanda Díaz-Merced, "Making Astronomy Accessible for the Visually Impaired," *Scientific American*, September 22, 2014.

4. Joan E. Solsman, "Internet Inventor: Make Tech Accessibility Better Already," CNET, April 10, 2017.

5. Andrew Liszewski, "Every Kid Can Enjoy a Day at the Waterpark with This Air-Powered Wheelchair," Gizmodo, April 2017.

6. Human Engineering Research Laboratories, University of Pittsburgh, "PheuChair Unveiled at Water Park," http://www.herl.pitt.edu/news-events/pneuchair-unveiled-water-park.

7. Cathy O'Neil, *Weapons of Math Destruction: How Big Data Increases Inequality and Threatens Democracy* (New York: Crown, 2016).

Chapter 9

1. John Gantz and David Reinsel, "The Digital Universe in 2020: Big Data, Bigger Digital Shadows, and Biggest Growth in the Far East," International Data Corporation, February 2013.

2. "Report on the OECD Framework for Inclusive Growth," May 2014.

3. Richard Samans, Jennifer Blake, Margareta Drzeniek Hanouz, and Gemma Corrigan, "The Inclusive Growth and Development Report," World Economic Forum, January 2017.

4. "The Future of Jobs," World Economic Forum Report, January 2016.

5. "The Future of Jobs," World Economic Forum Report, January 2016.

SUGGESTED READING

Chapter 1: Welcome

- *Inclusive*, a short film by Microsoft Design; www.mismatch.design.

Chapter 2: Shut In, Shut Out

- Vivian Gussin Paley, *You Can't Say You Can't Play* (Cambridge, MA: Harvard University Press, 1993).
- Inclusive: A Microsoft Design Toolkit, Subject Matter Expert Video Series, 2016, www.mismatch.design.

Chapter 3: The Cycle of Exclusion

For more information on the state of disability:

- "World Report on Disability," http://www.worldbank.org/en/topic/disability.
- UN's Convention on the Rights of Persons with Disabilities, http://www.un.org/disabilities/documents/convention/convoptprot-e.pdf.

Chapter 4: Inclusive Designers

For more information on the definition of accessibility and universal design:

- Sarah Horton and Whitney Quesenbery, *A Web for Everyone: Designing Accessible User Experiences* (Brooklyn, NY: Rosenfeld Media, 2013).
- Wendy Chisholm and Matt May, *Universal Design for Web Applications: Web Applications that Reach Everyone* (Sebastopol, CA: O'Reilly Media, 2008).

For more information on disability rights, identity, and history:

- Kim E. Nielsen, *A Disability History of the United States* (Boston: Beacon Press, 2012).
- Rosemarie Garland-Thomson, *Extraordinary Bodies: Figuring Physical Disability in American Culture and Literature* (New York: Columbia University Press, 1997).
- James I. Charlton, *Nothing About Us Without Us: Disability Oppression and Empowerment* (Berkeley: University of California Press, 1998).

Chapter 5: Matchmaking

For more information on urban design and social exclusion:

- Amy Arnold and Brian Conway, *Michigan Modern: Design that Shaped America* (Layton, UT: Gibbs Smith, 2016).
- Anthony Flint, *Wrestling with Moses: How Jane Jacobs Took On New York's Master Builder and Transformed the American City* (New York: Random House, 2009).
- Robert Caro, *The Power Broker: Robert Moses and the Fall of New York* (New York: Knopf, 1974).
- Claude M. Steele, *Whistling Vivaldi: How Stereotypes Affect Us and What We Can Do* (New York: W. W. Norton, 2011).

For more information on learning styles and gender inclusion in software design:

- GenderMag by Margaret Burnett, http://gendermag.org/.
- Marie Hicks, *Programmed Inequality: How Britain Discarded Women Technologists and Lost Its Edge in Computing* (Cambridge, MA: MIT Press, 2017).

Chapter 6: There's No Such Thing as Normal

For more information on the history of anthropological and mathematical ways of studying people:

- Margaret Mead, ed., *Cultural Patterns and Technical Change: A Manual* (Paris: UNESCO, 1953).
- Todd Rose, *The End of Average: How We Succeed in a World that Values Sameness* (New York: HarperOne, 2015).
- Clifford Geertz, *The Interpretation of Cultures: Selected Essays*, 3rd ed. (1973; New York: Basic Books, 2017)

For more information on cognitive inclusion:

- Steve Silberman, *Neurotribes: The Legacy of Autism and the Future of Neurodiversity* (New York: Penguin Random House, 2015).

Chapter 7: Love Stories and Outcomes

For more examples of inclusive design:

- Graham Pullin, *Design Meets Disability* (Cambridge, MA: MIT Press, 2009).

For more thoughts on the business case for inclusion:

- Mark Kaplan and Mason Donovan, *The Inclusion Dividend: Why Investing in Diversity and Inclusion Pays Off* (Brookline, MA: Bibliomotion, 2013).
- Vint Cerf, "The Internet Is for Everyone," speech to the Computers, Freedom and Privacy Conference, April 7, 1999, https://www.itu.int/ITU-D/ict/papers/witwatersrand/Vint%20Cerf.pdf.

Chapter 8: Inclusion Is Designing the Future

For more information on machine learning, artificial intelligence, and inclusion:

- Cathy O'Neil, *Weapons of Math Destruction: How Big Data Increases Inequality and Threatens Democracy* (New York: Crown, 2016).
- Sara Wachter-Boettcher, *Technically Wrong: Sexist Apps, Biased Algorithms, and Other Threats of Toxic Tech* (New York: W. W. Norton, 2017).
- Pedro Domingos, *The Master Algorithm: How the Quest for the Ultimate Learning Machine Will Remake Our World* (New York: Basic Books, 2015).

INDEX

SIMPLICITY: DESIGN, TECHNOLOGY, BUSINESS, LIFE

John Maeda, Editor

The Laws of Simplicity, John Maeda, 2006

The Plenitude: Creativity, Innovation, and Making Stuff, Rich Gold, 2007

Simulation and Its Discontents, Sherry Turkle, 2009

Redesigning Leadership, John Maeda, 2011

I'll Have What She's Having, Alex Bentley, Mark Earls, and Michael J. O'Brien, 2011

The Storm of Creativity, Kyna Leski, 2015

The Acceleration of Cultural Change: From Ancestors to Algorithms, R. Alexander Bentley and Michael J. O'Brien, 2017

Mismatch: How Inclusion Shapes Design, Kat Holmes, 2018